FARMER'S GLORY

Farmer's Glory

A. G. STREET

With decorations by
GWENDOLEN RAVERAT

faber and faber

This edition first published in 2009
by Faber and Faber Ltd
Bloomsbury House, 74–77 Great Russell Street
London WC1B 3DA

A CIP record for this book is available from the British Library

ISBN 978-0-571-25197-1

To
EDITH OLIVIER
who encouraged me
to write this book

AUTHOR'S NOTE

This book is simply an attempt to give a pen picture of farming life in Southern England and Western Canada. Whilst the characters are drawn in some measure from life, no names used refer to anyone in actual life bearing a similar name. Should any rural dweller consider that his or her particular type has been depicted unfairly, it has been done in honesty and without malice.

CONTENTS

FARMER'S GLORY

The hours of gold come back to me
That time has pinched (he can't return 'em),
The well-remembered chestnut tree
(Or was it, after all, laburnum?)
        The rural rill,
The shriek of dying pigs—I hear them still.

'Tis out of no bucolic whim
I promulgate agrarian measures;
But, now that Farmer's Glory's dim,
And Agriculture holds no pleasures,
        I'd give a lot
If I had farmed, when times like these were not.

With many apologies to *Punch*,
4th August 1909

INTRODUCTION

As I was turning out some shelves in the storeroom the other day, I came across one of my father's farm diaries dated 1906. The principal entries in it were the weekly wages paid to the farm labourers, but it also contained many notes as to cropping, stock, payments for shearing, threshing, and other entries connected with the working of the farm according to the season of the year.

For instance, in March there was a record of the number of lambs tailed, the payments made to the head shepherd, and also those to the tailer for his gruesome services. Later on in June came the piecework prices per acre for hoeing the various roots, and in August I found that Tom Scullard had mowed round one hundred and eighty acres of corn for one penny per acre. In these days it seems a lot to do for fifteen shillings, but I can remember Tom Scullard, and am sure that he would have been

11

INTRODUCTION

mortally offended if any other man had been detailed for the job, as he counted it as one of his privileges.

In the same month there was the curious entry of 'Hiling'. This referred to stooking the sheaves of corn. I am not quite certain in my mind as to whether it should have been spelt 'Aisleing', likening the rows of stooks to the aisles of a church, or whether 'Isleing' would have been more correct, as each stook is a separate isle of about ten sheaves. Possibly my father's spelling was entirely phonetic; anyway, it doesn't matter now, and it evidently served the purpose in those days, for I found that in that particular week six men had drawn eighteen shillings each on the 'Hiling'. Some weeks later I read that the balance coming to the men on the 'Hiling' was thirty-six shillings, which was paid to one, Charles Bailey; his faded signature sprawls right across the page.

I can remember him quite well: a short, thick-set peasant with a hatchet face. He was called Lawyer Charlie by his mates, as he always had charge of the financial side of their collective jobs. When the price of a piece of hoeing had to be settled, he was always the spokesman, and his great word in these discussions was 'feasible'. Apparently to single turnips at six shillings per acre was feasible, but the suggestion to do mangolds at anything less than seven shillings per acre always brought the retort: 'No, zur, bain't feasible.' Poor old Charlie is dead now. In my early days he was a good craftsman at all farm work, scrupulously honest, and he had a personal pride in the farm and everything and everybody connected with it. He was a chapel preacher on Sundays, and was greatly exercised in his mind because my father played cards. He would reprimand him for this deadly

12

sin in the middle of a discussion on farm matters, and then switch back to the business in hand very abruptly.

I heard him once say to my father: 'Devil's picture books, they be. You be headin' fur eternal damnation, zur. What did Christ say? Wash me and I shall be whiter than snow. Now about thic main carriage in Big Maid. . . .'

On looking through the list of names I found that quite a large proportion were not entered by their correct names, but by nicknames. 'Shiner' was one; his name was Frederick Strong. How or why he shone I cannot remember, but in that year he appears to have received a regular weekly wage of thirteen shillings plus his cottage, so he must have been a stockman of sorts.

Dick Turpin was another name on the pay-roll. This was a Thomas Trowbridge, who used to sing a song about Dick Turpin at the harvest suppers in the big barn. I have never heard it anywhere else; the last verse went something like this:

> *I am the last of Turpin's gang,*
> *And I am sure I shall be hanged*
> *Here's fifty pound before I die*
> *To gie Jack Ketch fur hanging I*
> *Tibby Hi Ho Turpin Hero*
> *Tibby Hi Ho Turpin Ho.*

Dick Turpin was a jolly old boy, an ex-shepherd who had become a day labourer, and helped with the flock at busy times. He found some of the piecework, especially hoeing, rather hard, as he had not been accustomed to it all his life. But he was always cheerful; and I was sorry on turning over the pages of the book to the date

13

of the Michaelmas settlings, to find this sinister entry: 'Received back from Dick Turpin five and sixpence overdrawn on hoeing.' Poor old Dick must have struck some bad pieces that summer, for I found that most of the hoers had a pound or two to come, and that Lawyer Charlie had four pounds eleven shillings and sixpence to his credit. But Dick was never a good hoer, and as he liked his pipe and glass I expect he had drawn more heavily than the others. Lawyer Charlie was a dour tee-totaller, and non-smoker, and as a boy I was a bit afraid of him, but I loved old Dick and wish he hadn't had to pay back that five and sixpence.

But what a number of names there were in those days; carters, shepherds, dairymen, labourers, a foreman, a groom-gardener, and boys and lads innumerable. Boys were rarely accorded the dignity of name in that pay-roll, as usually their fathers worked on the farm and drew the wages for the family. A man named Mussel-white had three sons working regularly. He was a head carter, and the entries went like this: Musselwhite 14s., ——9s., ——7s. 6d., and ——5s. In the harvest a boy still at school came to work, leading horses, and I saw that he was down as—Punch 2s. 6d. He was a very diminutive cherubic child with an uncanny knack with horses. To see him bring a load of sheaves hauled by two horses at length, into a stackyard through a gate-way, was to realize that there is a special Providence watching over children and dumb animals. He was rather a favourite of my father's, hence the entry 'Punch'; an ordinary boy would have been just ——.

On comparing that list of names with my present wages book in 1931, I find that the farm is now employing about 25 per cent to 30 per cent of the number of men. One or two of the younger men in that 1906 wage list are still with me on this same farm, but most of the old hands are dead and gone. Yet the perusing of that old diary brings back to my mind quite clearly memories of those days of my boyhood, and I cannot help thinking what a contrast the farm presents to-day, both in its appearance and personnel, as compared with a quarter century ago.

Whether that period may be truthfully described as the bad old days or as the good old days, I do not know. However, I have lived through that twenty-five years of farming, and possibly my reminiscences will show how the change occurred. If the present state of things be considered a bad one for this England of ours, perhaps someone will suggest a remedy. Why it took place I do not know; I only know that it happened. Therefore, in the words of Kipling, I will let a plain statement suffice, and leave you to draw your own conclusions.

PART I
THE SPACIOUS DAYS

CHAPTER I

My earliest recollections of this farm seem to be the holidays home from school. Possibly I took things very much for granted, but I can remember quite definitely that the farm was an extraordinary nice place to come back to from school. I don't know that I cared very much about the business side of farming in those days, but the pleasure side was another matter for a schoolboy.

The Christmas holidays were connected chiefly with shooting. My father got a lot of good shooting in the large wood at the top end of the farm. The landlord had large shooting parties of his own friends before Christmas, but after that rarely had a shoot without inviting some of his tenants. In January it was usually tenants only, and rabbits were the bulk of the bag, with a few cock pheasants and an occasional woodcock in addition.

I used to accompany my father on these expeditions, and was very proud of his skill. He was a fair shot at wing game, but at rabbits dodging between trees and undergrowth, or darting across a narrow track, I have yet to see his equal. He and his friends were always playing tricks on each other, rather like overgrown schoolboys. A favourite one was to plant a rabbit skin stuffed with straw down a rack where a notoriously bad shot would go in the next beat, in the hope that he would shoot at it; it was *infra dig* to shoot at a sitting rabbit. As usual the best humour was the unplanned, for on one occasion a very good shot, with a habit of boasting about the size of his personal bag, went down the rack by mistake, and shot at the stuffed rabbit to the great joy of the keepers and everybody else.

On rabbit days nearly every farmer would bring his own dog to help hunt, and it was necessary to shoot carefully, as there were dogs and rabbits running all over the place. One day, before lunch, someone shot a dog. Another individual, being a bit peppery, said after lunch that he wasn't going to have his dog shot, and accordingly tied it up to a bush. We went on without it—it was a very wild liver spaniel—and when we came back at the end of the day, we found that it had hung itself in its endeavours to join in the sport.

But the most enjoyable shooting party to me was a day's ferreting. Two or three under-keepers would come with their ferrets, and my father would invite two or three neighbours. The keeper of our beat was an old wizened man named Frank Hard. He would always bring his own gun, and I would wait eagerly, oh so eagerly, for him to say to my father: 'I don't want to shoot, sir, if young master Jim can manage my gun.' It was a very ancient hammer gun, but how proud I was to be entrusted with it. Throughout the day I was always placed in the most unlikely spot. 'One's guests must have the best places,' my father would say.

Presently the ferrets would lie up, the main party would move on to fresh ground, and I would be left with Ted Bridge to find the loose ferrets, and shoot anything that might bolt. Ted was a cheerful soul. He suffered from some form of bronchial catarrh, but he was a fine digger and drinker. He would cough, dig, spit, dig, swear, dig, and spit, in one continuous stream. On one never-to-be-forgotten occasion after a long dig, he pulled nine rabbits alive out of a hole, and put them in a bag. He then let them run one at a time for me to shoot, and I only let one get away. I was a very proud boy that day when we rejoined the main party, as Ted never breathed a word about the rabbits being first dug out.

As I have mentioned, he was a great drinker. He could drink beer without visibly swallowing; it just went straight down.

One day it was very cold, and he and I were sent ahead to a sheltered cover to build a fire, and put the lunch ready. He got a fire going nicely, and then tried the beer. Finding it too cold for his liking, he placed the jar close to the fire to warm it. When the party assembled for lunch it was found that the heat had cracked the jar, and that most of the beer was gone. Ted's face was a study.

One of the chief joys during the Easter holidays was connected with the water-keepers. At that season they were busy wiring pike. The essential thing for this sport was a still sunny day, so that there would be no ripple on the water, thus making it possible to spot the fish.

Long slender ash or hazel wands with a wire noose at one end were the implements used.

Pike lie absolutely motionless in the water until disturbed, and can be wired rather easily. At first it is a bit difficult to see whether your wire is round the fish or not, but you soon get the trick of it. Having spotted a pike within reach, you slide the wand gently into the water well above the fish, push it out to the estimated distance, and then let it drift gently downstream towards the pike. Sometimes you have not got it out quite far enough, and the far edge of the noose will scrape along the near side of the fish without disturbing it. When you judge it correctly, there are a few breathless moments as you watch your wire drift over the pike's head, down past the first fins, and then you give a joyous pull, feel the weight of him, and in a trice he is out on the bank, snapping his jaws ferociously.

The water-keeper who instructed me in this business was a gentle, lovable old man with a long beard. He is dead long since, but I shall never forget the happy days he gave me in the peaceful green Wiltshire meadows.

Organized shooting being finished at this season, there were still young rabbits to be shot. This was chiefly an evening occupation, and the weapon used was a .22 rook rifle. My father was carefulness itself in the handling of a gun, and would allow no liberties to be taken by anyone in his company. The earliest smacking I can remember was administered by him in the nursery, because I had pointed a toy gun at the nursemaid. He had a great idea, too, of teaching children to connect cause and effect. The first time I accompanied him on one of these rabbit-stalking expeditions with a rifle, he said I could come if I would keep quiet. I agreed to do so, but was soon chattering gaily. During our walk he told me about it several times to no purpose. We got no rabbits that night. A few evenings later he took the rifle down from the gun rack, and I immediately asked if I could go. 'No,' he said. 'I want to get some rabbits. You talk too much.' I remained at home, very miserable, but the lesson was learnt for all time, and what fun I had with him later on those expeditions, creeping, sometimes crawling, being especially careful never to put one's foot on a stone or rotten twig, and, above all, learning to shoot with a rifle not only quickly but safely, always remembering that a rifle bullet goes a long way.

I suppose boys still use catapults, but in those days I rarely went out of doors without this silent and efficient weapon. It has one peculiar joy in that you can watch the bullet in its flight until it strikes or misses the mark. The first victim of my skill was a sparrow on the cowhouse roof, which I slew with a small potato. I can feel it between my fingers now, still see the glorious curve of its flight finishing square on the sparrow's chest, and sweeping him into oblivion from the ridge into the cowyard on the other side of the building. And I ate him that night for supper, after roasting him on a string in front of the kitchen fire. Looking back on those days I do not think that I failed to slay a specimen of every species of bird and animal life on the farm with this weapon, save only the lordly heron and the elusive snipe.

August holidays brought the joys of the harvest, both pleasure and toil. The main sport was the killing of the rabbits in the corn fields. I believe rabbits come under the heading of vermin, which is not a nice name to give to an animal which affords such a lot of pleasure to bloodthirsty boys. My father would stand with his gun between the burrows and the remaining piece of standing corn. The village boys and I had to keep back out of his way, but were at liberty to chase any rabbit that ran out on our side of the field. It sounds very cruel perhaps, but in those days to run a rabbit down, fall on it, and then slay it with one's bare hands, was a gorgeous experience.

Helping with the harvest work, and, as one grew older, being promoted from the lowly leading of horses to the final peak of rick-making, was a satisfying pleasurable thing. Also, for this, I was paid in cash every Saturday with the men. More cause and effect.

This record seems to be all about pleasure and sport. I suppose the business side of farming had its worries in those days, but it is difficult to recall any. There were good seasons and bad ones, doubtless. I can remember wet weather in harvest time and good weather. Good luck at lambing time and bad I can also call to mind, but nothing ever seemed to make any difference in our home life. It all seemed such a settled, certain, prosperous thing. Not that there was any display of luxury or any ostentation of any kind, but there was a spaciousness and an aura of solid well-being that is sadly lacking today.

CHAPTER II

I left school in 1907, when I was between sixteen and seventeen, and started in to the serious side of farming. The arrangement was that I should receive my board and lodging at home, and seven shillings weekly in cash, upon which I was to dress myself. It is a very long time now since I saved any money, but during that first year, I remember, I put away five pounds, in addition to dressing myself to my own complete adolescent satisfaction.

Frankly, I cannot remember doing much actual hard manual work except in rush periods such as lambing, haymaking, or harvest. There was such a crowd of men about the place. Thirteen horses were kept; six ploughing teams, and an odd horse to wait upon the sheep. This meant seven carters. Sixty dairy cows necessitated six milkers. The Hampshire Down flock of four hundred ewes required two shepherds. Six day labourers were employed, a foreman, and a groom-gardener. A grand total of twenty-three employees all the year round, with many more at harvest and haymaking. There were four hundred acres of arable land, ninety acres of down, forty acres of water meadow, and about a hundred acres of pasture.

I don't think my father can have been an average type of tenant farmer, or perhaps he worked it out on the theory that 'one doesn't keep a dog and do the barking oneself'. What I mean is that he didn't get up at an unearthly hour every morning. But I did. I was his dog. But even so he didn't forget to bark when occasion required it. When any of his neighbours chaffed him about lying in bed, he would retort that when he was up he was awake. He would get up early for anything definite: to go cub-hunting, shooting, to a fair, at harvest time, or other important occasions, and lastly, when none of us expected him to do so. I realize now that in this last lay the secret of his success. This, of course, is rank heresy according to accepted agricultural standards, both ancient and modern, but I submit that no text-book on farming should omit its possibilities.

I had been to an agricultural school, where we were taught that unremitting personal attention to detail, from early morn to dewy eve and after, was the first essential to successful farming. It was a bit of a shock, therefore, to come in about 9 a.m. to report, and to find the successful farmer in bed. I have inherited his liking for bed, but not, unfortunately, his faculty for making money out of farming.

THE SPACIOUS DAYS

In those early days my chief duty was that early morning round of inspection with its ensuing detailed report to my father. Afterwards the 'organizer', as one or two of his old and trusted hands called him behind his back, took charge of affairs, and things happened; chiefly because they were necessary, and sometimes, I think now, for moral effect.

The usual procedure was for me to meet the foreman at 6.30 a.m. We studied the weather, and planned the work for the horse teams. When I say we, in reality he did most of the studying and all the planning. After I had been home a few months I began to make tentative suggestions, which were never approved. Probably they were mostly wrong ones, but anyway, is a man of sixty who started work on a farm at ten years of age going to listen to a young whippersnapper of seventeen, fresh from an agricultural school? I ask you? Mind you, he was very nice about it. He always treated me with a subtle deference as the young guvnor, especially in front of the other men, but all the time I knew, and so did he.

The carters came to harness their horses at a quarter to seven, and, having given them their orders, we were ready to detail the six day labourers to various jobs at seven o'clock. The foreman had absolute authority over the carters and labourers, but the head dairyman and head shepherd were in a class apart; they had charge of their respective departments, and the men employed under them.

My next job was to visit the sheepfold, and find out if all was well. I feel that in writing this next sentence I am giving away another bit of invaluable information which the student of farming will never find in any text-book. 'Shepherding qualifications being almost equal, choose a cheerful shepherd.' There are so many vicissitudes which can happen to a Hampshire Down sheep, from the cradle to the grave, so to speak—I should have written table instead of grave—that a cheerful shepherd is the only type possible to overcome them. I have known many shepherds, and say with authority that the mournful ones are beaten from the start.

From the sheep I would go to where the horses were at work to see if everything was working according to plan, then back to the farm buildings to have a word with the head dairyman, and finally in to breakfast, usually a substantial meal.

Sometimes my father received my report at breakfast, and sometimes, as I have said, in his bedroom. This done, he rapidly decided his plan of campaign for the day. He was a bit crippled with rheumatism even in those days, and used to journey about the farm in a governess cart. The pony, Tommy, had a mouth like iron, an insatiable appetite, and the happy knack of trotting at a slower speed than he walked. Early in their acquaintance Tommy had discovered that as long as he trotted my father was content, and he had developed this slow trot to a fine pitch of perfection. On most mornings I would be instructed to tell the groom-gardener to bring the trap round at ten o'clock; there was always somebody to tell in those days.

The farming was on a settled definite system, the result of centuries of experience. The arable land was divided into four fields of one hundred acres each, and was cropped as follows:

THE SPACIOUS DAYS

First year. Winter-sown corn, either wheat or winter oats.

Second year. Half in clover for hay, and the other half into rye, winter barley, and vetches for spring sheep keep, followed by swedes and kale for winter feeding. The clover was alternated on to the other half every four years, as clover will only grow successfully in this district once in eight years.

Fourth year. Summer roots, usually rape and turnips.

This rotation was as unalterable as the law of the Medes and Persians. One always knew what crop a particular field would be growing two or three years ahead, and worked to that end. Any slight variation was considered a sin, and, like sin, it always left its mark. For instance, if one were tempted—I use the words advisedly—to seed a piece of vetches or clover, the extra robbing of the ground showed in the ensuing wheat crop. It mattered not a whit that the produce of this immoral seeding might bring in more money than a good crop of wheat. One didn't farm for cash profits, but did one's duty by the land.

This was one of the chief reasons for the inherent conservatism and mistrust of new things so prevalent at that date amongst the agricultural fraternity, both masters and men. If any new method were tried, one didn't look for its advantages, one ignored them, but one missed no opportunity to point out its defects. My father was quite keen on new things, partly, I think, because the foreman was always dead against them. That worthy would usually finish up the argument with 'Doan't 'ee do it, zur. 'Tis wrong.' Between them they adopted new things rather quicker than the average.

29

This same four-course rotation is still practised religiously here and there in Wiltshire and the south of England to-day, chiefly by men who have plenty of money. And they, good farmers all, are watching their capital shrink steadily, year after year, and they do so in a hurt and bewildered frame of mind. They are farming honestly and well, and losing five pounds per acre on every acre of corn they grow. They carry on year after year in the hope that the turn of the tide will come, but one by one they are forced either to drop out of farming altogether, or to adopt other methods.

Under the rotation mentioned it is easy to see that half the arable land was put into corn each year, and the other half devoted to growing feed for the flock, whose continuous and regular folding over the land made the corn growing possible. It always seemed to me that the farm was run entirely for the sheep, and most of the men were jealous of the shepherd's consequent importance. One of my father's labourers frankly hated sheep. 'All we do do,' he would say to me, 'is run about and sweat after they blasted sheep. We be either lambing 'em, runnin' 'em, marken 'em, shearing 'em, dipping 'em, or some other foolishness. And they can have all the grub we do grow, and God knows how much it do cost the Guvnor fer cake.'

I do not think we were very special with sheep; only about average. It was a pedigree Hampshire Down Flock, but we never went in for cups and shows. From my own experience I have come to the conclusion that there were two ways of running a Hampshire flock successfully in those days. One was to give them whatever quantity of cake they would eat, and to consider them and to worry over them far more than one did over one's own children. The other was to do them very badly, avoiding prosecution by the R.S.P.C.A. by the smallest possible margin. The middle course, which my father adopted, did not pay much, if at all.

But sheep are annoying things, and so are a good many shepherds. During a barren late spring, when you were short of grub, most shepherds would delight in seeing how fast they could gallop over it, pitching out larger and larger folds each day, and never grumbling about the extra work. Given a plenteous season, when you wanted the keep cleared faster before it spoiled, the shepherds would feed it in a niggardly fashion. Any suggestion as to speeding up was greeted with the definite remark: 'No, zur, t'wun't do.' From this there was no appeal. The shepherd's word was law. The rest of us just grumbled and carried on.

I don't know whether some shepherds will prosecute me for libel over this, but as a general rule, save for lambing, and other busy times, a shepherd reckoned to finish his actual laborious work by dinner time. After that he studied your sheep. I remember a shopkeeper in a neighbouring town, who retired from business at fifty, and took a farm in this district. Early in his rural career he went out one afternoon, and discovered his shepherd dozing under a bush on the down, with the flock grazing around him. 'What in the world are you doing, shepherd?' asked his employer. 'Lookin' atter your sheep,' replied the shepherd.

'Yes, yes, that's all very well, but you mustn't sit down. I can't pay you to sleep. You must get up and cut thistles, chop down some of these bushes, or do something.'

'Well, I bain't gwaine to. I be studyin' your interests, I tell 'ee, same as I allus have fer any maister.'

The farmer in question told my father about it afterwards. 'When I think of how I used to run up and down behind my counter, it makes my blood boil,' he said.

Both master and man were right in their judgment of the situation, and afterwards had a sound mutual respect for each other. But shepherds were always studying their sheep, and never seemed to tire of it. It always used to amaze me at our harvest suppers, where we could have a choice of beef, mutton, or ham, that our shepherd always chose mutton in large and repeated helpings. Whether he did so with the idea of supporting his own industry, or in order to get his own back on one of the animals who ruled his whole life, I do not know, but it was always mutton for him.

This autocracy of shepherds will sound almost unbelievable to townsfolk, I expect, but it was a very real thing in those days. Another instance of it comes to my mind. One of the largest and most successful rambreeders in this district discovered one season that he had a large rick of good hay untouched. He farmed near a racing stable, the owner of which offered him a fancy price for the rick. He sold it, and a few days afterwards mentioned the fact to his shepherd in conversation. The shepherd ruminated for a few moments in silence. 'Oh, you've selled un, 'ave 'ee?'

'Yes, what about it? We don't want it.'

'Not thease year, p'raps, but I reckons to be consulted about a thing like that. Still, 'tis yourn, to do as you likes with. But I shall leave.'

His master paid the buyer ten pounds to cry off the deal.

Usually the dairymen were of a different type from the other farm men. Producing milk twice daily, seven days a week, and railing the product daily savours more of a factory than a farm, and has its consequent effect on the men engaged in it. Even in those days they had a shrewd suspicion that the milk found the money which ran the whole farm, and their scorn of sheep and shepherds was very bitter. And they ran their job without outside assistance. Give them cattle and food, and they would do the rest, literally producing the goods.

I do not think that my father knew how many cows he had, to one or two, and am certain that he had no intimate knowledge as to their different milking qualities. The milk was sold on contract to London, with a minimum and maximum daily quantity. When the daily output went down dangerously near the minimum, or the cake supply was getting low, the head dairyman would come over to the house, the day before market, and ask to see my father. 'I dunno whether you knows it or no, but we be gettin' short o' milk. Thee better get two or dree heifers in market.' Or possibly: 'We be nearly out o' cake, and you'd better get the next lot kibbled, as we be main busy just now.'

This news was like the sight of hounds to an old hunter, and my father would set off for market next day, thirsting for the fray of buying and selling, and taking me with him for educational purposes in this most difficult art.

The cows were kept in two lots. One was the main milking herd, and the other, a smaller lot, consisted of

33

the dry cows waiting to calve and the cows nearly dry. This latter herd was in charge of a dear old dairyman, who, thank heaven, is still with me as I write. He and his wife ran this herd between them, and the entire management was in his hands. My father on his rounds in the trap would drive into the yard at these buildings, and shout, and if this was not successful, bellow: 'Strong?' From the depths of the building would come 'Yes, zur.' Another hail: 'All right, Strong?' Again would come the 'Yes, zur.' This intricate business being concluded, my father would drive away, having satisfactorily seen to that dairy for the day.

I am afraid all this will be rather apt to give the impression that my father was a bad farmer, and in the hands of his men. He may have been, but they never knew. They respected him and loved him, and he respected and relied on them. But this is the acid test. His methods paid, and paid well. When I think of the worries of farming now as compared with those simple days, I have a great admiration for that period and the men engaged in farming at that time. With all the rationing, recording, and worrying over each individual cow that I have done since: with all the scientific methods of farming in all its branches that I have tried; and with all the work and intense personal attention to detail that I have put into every department of my farming, I am forced to admit that my father was a much better farmer than I. He made it pay, and I have lost more money than I care to think about.

CHAPTER III

The point I would again stress about this type of pre-war farming is that one didn't consider whether the crop one was sowing would pay a profit over the cost of production or not. That never entered anyone's head. In good seasons farmers did pretty well, and in bad ones, presumably not quite so well. Granted, there were occasional instances of farmers going bankrupt, but these rare cases could always be traced definitely to drink, gambling, or some other vice or extravagance causing neglect of the farm by the master. If one attended to one's business decently, one got along all right. Some did better than others, but all got along all right.

The fact was that the four-course system allied to a Hampshire Down flock paid pretty well in those days, and was the accepted practice of the district. Farms were laid out for it, and let on the understanding that the customary rotation would be followed. And once you were fairly into that system, it swept you with it, round and round, year after year, like a cog in a machine, whether you liked it or not.

Let us take the start of the farming year at Michaelmas. The cleared stubble of the previous spring corn that was not sown to clover had to be ploughed and sown to rye, winter barley, and vetches, to provide sheep keep for the following May, June and July. It was no good wondering whether this would pay. You had four hundred ewes, which would, God willing, produce some four hundred lambs in January, and require food in May, June and July. In October, these ewes would be folding off rape and turnips, and behind them would come your ploughs and harrows, sowing this land to wheat. But would wheat pay? What were the prospects of the world's next harvest? Don't be silly! This land was due to come into wheat, and wheat must be in by the end of November if possible, so you didn't worry over abstruse problems, but got down to the job.

This done, the wheat stubble was ploughed up by Christmas, cross-ploughed in January, and sown to barley and oats in February and March. By this time your flock would have lambed, and be folding off swedes and kale. You have only got enough keep sown to keep them until the end of July, the rye, etc., which you put in in October. They will want rape and turnips from then onwards, so your horses must follow them in April, as they feed the swedes and kale, and sow rape and turnips.

But what about next winter? You will require swedes and kale for the flock. So your ploughs and harrows must follow them in May, June and July, as they feed the rye, winter barley, and vetches, and sow swedes and kale for next winter.

36

And so it went on, year after year, one continual hopeless striving to feed the flock. Sheep! Sheep! It was always the sheep. Your life was ruled by them, the whole farm revolved round them, and, in my case, my father's temper varied with the state of the flock's well-being. They were a kind of Moloch, to which we were all sacrificed.

The old labourer referred to in the preceding chapter was quite right in saying that all we did was to wait upon their needs. He used to vent his hatred on them, when they were dipped. He would stand at the side of the swimming bath filled with Cooper's Dip, armed with a long pole with a cross piece at the end, and push them under the evil-smelling liquid with great glee.

We got a little respite from this eternal striving to satisfy the sheep in August and September, when the whole farm concentrated on the harvest. Sometimes even the shepherd would help in the evenings in a condescending sort of way, as one conferring a favour. I generally used to take my dinner with me during harvest. Field work started at 6 a.m. and continued till 8 p.m., and it was much nicer to eat bread and cheese and cold bread pudding in the field, than to waste the dinner hour in the journey home and back.

Many people think that the agricultural labourer of those days was slow in his movements. This is incorrect. He looked slow, I grant you, but the experience of countless ages had discovered the simplest and easiest way of doing his manifold laborious tasks, and years of practice had transformed his gnarled and clumsy hands into extraordinarily deft and dexterous instruments. Also, he struck a gait at any job which he could keep up from daylight to dark, day after day. Any attempt to hurry him was disastrous. He considered it to be a slight on him, and that you didn't realize that he always gave of his best.

When we were carrying corn the number of pitchers in the field loading the wagons, and the number of men at the rick emptying them, had to be regulated to a nicety to keep the wagons going backwards and forwards steadily without a hitch. That is where Tommy and the trap came in. Tommy would be required earlier during the harvest; as a matter of fact he stayed harnessed to the trap all day until knock-off time. The wagons were led from the field to the rick and back by small boys. One of my father's dictums was that two boys together did half as much as one boy by himself, and that three boys did nothing at all. Let two of the boys stop for a minute or two as they passed, one with a full wagon and the other with an empty one, and the whole business of carrying was disorganized. Then, from some point of vantage, Tommy and the trap descended on them like the wrath of God. On these urgent occasions Thomas moved quite smartly.

I can appreciate now that my father's work in this way was very important. He would also play off the rick staff against the field men. Perhaps we would be a pitcher short of the required number owing to one of the carters having gone to the station for a load of something. Father would drive out to the other pitchers, and say to the head carter: ' 'Fraid you won't be able to keep us going so well to-day till Fred gets back. Still, we must put up with it. Just do the best you can.' Or perhaps the rick staff would be a man or two short for a similar reason, and it would be: 'I wonder if I'd better drive down to

the dairy, and see if a milker can be spared for a bit·
You chaps won't keep those pitchers going else.' 'Doan't
'ee worry, zur,' they would be sure to say, 'we'll manage
somehow.'

If the occasion were desperate, and another hand must
be got somehow, the shepherd was the last resort. You
didn't send the foreman to see if the shepherd could get
away for an hour or two. That would have been to court
disaster. The sheep would have been in such a critical
state that if the shepherd left for a moment, they would
all be sure to die. Neither was I sent. Youth hadn't the
tact required for such a ticklish operation. Oh no! That
was a job for the Guvnor, and from the rick we would
see Tommy being urged to his most furious speed up the
far slope towards the sheepfold. Having arrived, my
father talked sheep, sheep, and nothing but sheep, thus
relegating the harvest to an unimportant detail unworthy of mention. After a bit the shepherd would be
sure to say: 'And how be getten on wi' the carrying,
zur?' 'Pretty fair, shepherd. We're a bit shorthanded to-
day. I'm on my way down to the village to see if I can
pick up another man.' 'Well, zur, I be about straight yer
just now, in a manner o' speaking. Ud it be any good if
I were to gie 'ee a hand fer an hour or two?' And back
to the harvest field would come Tommy, hauling both
the shepherd and my father, who had achieved his object without mentioning it. All this may sound childish
to many people, but some will, I hope, recognize it for
what it undoubtedly was—pure genius.

My own personal relations with the men in those days
were of the best. I know that I liked them, and I think
that they liked me. I did not do much actual laborious

work, but my father made me do every job on the farm at some time or another in order that I might, from personal knowledge, be able to estimate whether a man was working well or ill at any particular job. I was much older before I realized how much I did learn in those first years after leaving school. It is curious that one doesn't know the exact moment when one felt qualified to say whether the sheep were doing well or not, whether a certain cow or horse were a good or bad one, or what precise cultivation was needed for a particular crop or field. One only knows that suddenly one does know. You don't learn by going round and asking why, but by growing up with the whole business. One assimilates knowledge unknowingly. And it isn't all knowledge. One learns a good bit about faith in the beneficent wisdom of the Supreme Being. I remember once the foreman saying to my father when he was going to carry some barley rather quickly after cutting it: 'Don't 'ee do it. Thee's left it to the Almighty fer six months. Let un have it dree more days.'

Naturally I learnt a lot from the men. You cannot work with people day after day, all the year round, without doing so. They were, without exception, very definite about things. They never explained anything. They just said so. And usually I found out that these statements were correct. Our old dairyman once said to me: ' 'Tis no good buying cattle from down stream fer these meadows. They don't do up yer. You wants to goo up stream and get 'em.' Ten years afterwards I bought a lot of cattle from down stream, and they didn't do. That deal cost me at least a hundred pounds, and then I appreciated the old man's words.

They didn't mince matters, either. The first harvest after I left school, I was one day on the top of a rick with an old rick-maker. The surface we were working on was getting smaller and smaller as the roof drew in, and I was being particularly careful not to get in his way, and to place each and every sheaf just to his hand, when he looked up and said: 'Thee best get down.' I got down.

There is no doubt that the agricultural labourer is much better off now than he was during the period of which I am writing. He has to-day a higher standard of living, a broader outlook on life, and a taste for amusements and interests outside agriculture, but whether he is any happier or more contented is open to question. Definitely he is not such a good farm hand. These other interests distract his attention from the farm. I do not say this in any spirit of criticism, but merely state the fact. Why should he worry about the farm after his working hours, allotted by law, are finished? But twenty-five years ago his sole interest was the farm on which he worked. Nowadays he does what he is paid for, but then he did what was right and necessary to the well-being of crops or stock irrespective of payment. Then he took a pride in his particular department of the farm, and also took the responsibility of it, but now he runs to the boss for instructions at every touch and turn. The same alteration in outlook and amusements has taken place in the farmers also, with, I think, the same consequent deterioration in their value to the land. In those times, the farmer's sole interest was his farm. What went on in the world outside of farming he didn't know, and didn't care. The farm supplied all his amusements also: shooting, hunting, fishing, local tennis—one didn't go to Wimbledon in those days to watch tennis, one concentrated on the best method of dealing with one's neighbour's devastating first service, a very real and urgent problem—an occasional point-to-point meeting, a puppy show, and countless other festivities pertaining to one's calling. One never got away for a moment from the atmosphere of farming. Both farmers and labourers might have been justly called narrow-minded clods by townsmen in those days, but as guardians of the soil in their particular district they were unbeatable.

But that large tenant farmers were doing pretty well then, there is no question. As I have said, they did their duty by the soil, and, in the words of the Scriptures, it repaid them, some twenty, some sixty, and some a hundredfold. Home scenes of that period come back to me quite clearly. A tennis party, some ten or a dozen people on the lawn, a man and a boy bedding out geraniums and lobelia, and in the background the rambler roses flaming in all their June beauty. The local hunt running a fox to ground not far from the farmhouse, and Tommy and the trap being immediately requisitioned to take liquid refreshment to horsemen and diggers. The meet of a shooting party at nine o'clock on a December morning: friends and neighbours arriving in governess cars and dogcarts; the varied fortunes of the day's sport and the gargantuan high tea in the evening, which was invariably followed by cards till all hours. Not the quiet solemn bridge of to-day, but the blunt, noisy, cheery penny nap, followed by an hour of that full-blooded gamble, Farmer's Glory, to wind up the evening. I believe the correct name for this game in urban society is

Slippery Ann, but that Farmer's Glory was the more appropriate in rural circles there can be no disputing. The whole business of being a farmer in those days was indeed a splendid glory.

No record of this period would be complete without a mention of the weekly pay-day. What a business it was! The yellow canvas bag had a partition dividing the gold from the silver, for we had gold in those days. Most farmers paid fortnightly, but my father always said that the money was of more value to the men if they got it weekly, and paid accordingly. He made a special point of going round the farm and paying the men at their work. There were no time-sheets. 'What do you make it this week, Tom?' my father would ask one of the labourers. 'Two and a half days fer you, zur, four hours overtime, and t'other hoeing.' 'What'll you draw on the hoeing?' The hoeing was piecework, and the 'two and a half days fer you' day work. 'Aight shillin', zur, and I've a ketched two dozen moles.' And from his pocket Tom would produce a grimy screw of paper containing twenty-four moles' tails, for which he would be paid one penny each.

We met the head coachman of the estate during the round one pay-day, and after due salutations my father offered him the screw of paper containing the tails for twopence, on the ground that he had purchased it for two shillings only five minutes previously. I was only about fourteen at the time, and I can remember wriggling with suppressed excitement, as the coachman rode up to the trap and eyed the parcel suspiciously. 'Well,' he said to my father, 'if you've paid two bob for it, I don't mind chancing twopence.' Which he did, to my father's great satisfaction and my own undisguised glee.

On the pay-day before Christmas, the men received Christmas boxes, carefully graduated according to their proper status. On these occasions the foreman, head shepherd, head dairyman, and head carter, received orders on the local butcher for joints of beef, while the other hands got cash, half-crowns for men, down to sixpences for boys. It was indeed a perfect example of that ideal form of government, a benevolent autocracy.

CHAPTER IV

Harvest finished usually early in September, and the final event in the farm year was the large sheep fair in the middle of that month. The only work done on Fair day, apart from essentials such as milking, was the driving of the three hundred odd sale lambs to the fair. Half the lambs would be ewe or chilver lambs, and the best one hundred and fifty of these would be kept at home to replenish the flock. The male or wether lambs would be sorted into a best hundred and two fifties. The remaining ewe lambs would provide another lot of fifty, and the odd ones of each sex would be put together and sold as so many mixed lambs.

As the farm was only a bare two miles from the fair ground, the sale sheep were driven to the fair in separate small flocks, as they were to be sold. Truly, a lordly procession, which left the fold at 4 a.m., led by the head shepherd and the best hundred lambs.

I have seen seventy thousand sheep penned and sold in that fair in one day, in addition to many cattle and horses. To the uninitiated the fair field must have appeared as a confused muddle of rustics, sheep, and dogs, with a cacophony of baas and barks as a never-ending accompaniment.

I think that the auctioneers who sold the sheep enjoyed the day's work thoroughly. I can see the leading one now. He had a raised plank walk down between the pens, from which he surveyed the sheep and company as an admiral upon his quarter-deck. Servant of the farmers he might be, but when selling that day, he was master of all, both buyers and sellers.

The sheep sold, the farm settled down to the customary ploughing and planting for the next season, but before I leave this prosperous period there was one more happening in the farming year which is worthy of mention. I refer to the harvest supper.

The big barn would be cleared out sufficiently to make room for the seating of thirty or forty people. The words, 'God Speed the Plough', in letters eighteen inches high, on the wall at one end, would be freshened up with whitewash; this was usually my job. A local caterer would arrive in the morning with tables, benches and provender. The menu consisted of cold meat, beef, mutton, or ham, with hot boiled potatoes for the first course, and hot figgety duff puddings with whole raisins in them for the second. They were boiled all the day in the dairy copper.

There were no windows in the barn, and the lighting

was provided by oil lamps hanging from the beams on chain plough traces. I can visualize that scene quite clearly: three tables in U-shaped formation, my father in the chair at the top table, and the foreman and myself at the ends of the others. I can see the ruddy countenances of the company, shining like burnished copper in the pool of light from the lamps overhead. The light would filter through cracks in the reflectors here and there, and faintly outline the arching rafters of the barn, giving the whole scene almost a churchlike appearance, Indeed, the company might have been a gathering of old and jolly friars, save that whiskers predominated.

Most of them drank beer, and here I would like to correct the common impression that the farm worker of that day lived only for beer. Very rarely did I ever notice anyone drink to excess on these occasions, although a certain mellowing was apparent. And there was no stint. A barrel was horsed at one end, and they helped themselves. As a class they had arrived at the correct valuation of beer in their scheme of things. For long hours of manual labour in hot weather, beer was the best drink. In the days years before, when part of their wages consisted of beer, the allowance for a carter in the harvest field was one gallon daily, while the head carter was allowed five quarts. And in my day in the harvest, if you wanted to get a field finished before night-fall, you were far more likely to get the extra effort out of your men by sending out some beer in the evening than by any promise of overtime pay. But on an occasion like this, to get fuddled with beer when you were not working, was deemed a breach of good manners, and frowned on by all.

THE SPACIOUS DAYS

Supper finished, and grace said, the toast of the King was given, and then china basins filled with shag were placed at intervals down the tables. The effeminate, who smoked only cigarettes, had to provide their own. I was in this category in those days, and used to take a plentiful supply with me. My father would beckon me to take his cigar-case round to the foreman and heads of departments. There did not seem to be any jealousy at this subtle distinction. After all, the farm labourer lived so close to nature that he had a true sense of fundamentals, and thought it only seemly to give to every man his proper due according to his station. Also the non-cigar men got a good deal of amusement out of the laborious efforts of the favoured ones to smoke their cigars. The shepherd, I remembered, burnt his at one end and chewed it fiercely at the other. When you consider that he had the best part of a leg of mutton inside him, it was probably wisest to finish the cigar as quickly as possible. As his face was whiskers all over, I used to imagine horrible catastrophes as the glowing end got nearer and nearer to the jungle.

The evening then developed into a smoking concert, interspersed with speeches, much like a lodge dinner. First, my father would give the toast of the staff. I cannot remember the words he used, but he sincerely thanked them for their year's work, and gave them the impression that it was greatly appreciated by him, which it was, and he would finish up with a special reference to the foreman, the head shepherd, the head dairyman and the chief carter.

The visitors and he and I would then rise and solemnly drink their health. Then my father would ask if anyone would give us a song. After much chaff, one of the under-carters would drink deeply, and rise to his feet prepared to do his worst. Their songs were chiefly of a morbid sentimental type. Some I can remember now. 'It was only a beautiful picture in a beautiful golden frame', was one; something about angels' eyes watching a lonely little cabin on the railway line and averting some terrible catastrophe, was another; and 'They laid him sad and lonely in his faded coat of blue' was a third. A note of humour crept in when one of the labourers sang about the delights of jumping out of bed when the cock begins to crow, as my father's love of bed was common knowledge.

However, when once started there was no need to call for volunteers; they were ready and waiting—at least amongst the younger men—to exhibit their prowess. The old men bided their time. 'Let these boys finish their squawking, and then we'll show 'em zummat', was their attitude. Once, I remember, we got a modern note. A young dairy lad about sixteen sang 'All the nice girls love a sailor', which some of the old hands thought hardly suitable for the occasion. That particular lad is now managing a large Grade A Tuberculin Tested Dairy up in Hampshire. Not only in his choice of song at sixteen was he more up to date than his fellows, but also in the outlook which shaped his subsequent career.

However, his song brought a lull in the proceedings, and the foreman seized the opportunity to rise to his feet and call for order. His speech, as I remember it, went something like this. 'Chaps, we be all yer once again, and we be glad to be yer, and wishes to say thank you to the Guvnor fer a downright good spread. We've a finished another year together, and we be truly thankful. The Guvnor, he said as 'ow we ain't 'ad no onpleasantness, and that 'ee do vally our work. Well, speaken fer one and all, I says as 'ow we do vally to work fer un. Course, there mid a bin a bit ov a miff, now and agen. The Guvnor, 'ee will get hisself into sich a tear at haymakin' and harvest. But there, we do know as a can't 'elp it, and we've a come droo thease year all right, and shall agen, please God. Still, 'ee do play fair, in the manner o' speaken, and we do vally it. We be pleased and proud to work fer un. Course, there's the young Guvnor. 'Ee do know a lot, and there's a limb ov a lot as 'ee don't know. But 'ee don't do so bad, and we be a learnin' 'im smartish. But I low fore I asks 'ee all to drink their health, that I muddn' ferget the Guvnor's wife, who dod do a lot fer we. You do know, chaps. When you be ill, or when yer wife do 'ave a youngster, 'tis the Missus who do make it go easier and more suent like. Well, now then, let's 'ave you. We'll drink their health and long life to 'em all.' The toast over, the foreman would start the chorus 'For they are jolly good people', to which an accompaniment would be played on an accordion by the head carter.

The serious business of the evening being over, the old dairyman would offer to give us a song. He was, and is, a dear gentle old man, and he had usually two or three goes before he could pitch his song so that he could sing it without breaking down. After two or three false starts he would have another swig at his glass, beam at the company, and launch forth into that old South Country song, 'Buttercup Joe'. As the singer is happily still with me as I write, I am able to give you the words of his song. I asked him the other day if he would write them out on a piece of paper for me if he could remember them, and he agreed to do so. This morning, I was over in the dairy, printing up some butter, and he came up to me and said: 'About thic zong, zur. I bain't much ov a scholard like, and they at whoam don't want to bother wi' un. But we be yer by ourselves now. There iddn nobody about. If I were to say un to 'ee, you could write un down, couldn 'ee?' So on a sheet of butter paper I wrote out the song as he repeated it. He did so in a sing-song voice, and kept straying into the tune at intervals. Here it is.

BUTTERCUP JOE

1

O, I'm a breastin sort ov a chap
Me father comes from Shareham
Me mother got some more like I
She well knows 'ow to rear 'em
O, some they call me bacon vace
An' others turmut 'ead
But I'm as clever as other volk
Although I'm country bred.

Chorus:

I can drive a plough
Or milk a cow
O, I can reap or sow
I'm as fresh as the daisies in the fields
And they calls I Buttercup Joe.

THE SPACIOUS DAYS

2

You 'eavy swells mid laugh and chat
To see us eat vat bacon
But you can't drink our country beer
And that's where you're mistaken
A drop o' moey and shannon too
You drink it at your ease
But give to me an 'omebrewed glass
With crust ov bread and cheese.

3

O ain't it prime in summer time
When we go out haymakin'
The lads and lasses with us chaps
Freedom will be taken
And don't they jiggle and make us laugh
Ov course in harmless play
They likes to get us country chaps
To roll 'em in the hay.

4

O, do you know my young ooman
They calls her our Mary
She works as busy as a bee
In Farmer Johnson's dairy
And ain't her suety dumplins nice
My gosh I mean to try
And ask her if she'd like to wed
A rusty chap like I.

The spelling of the song is entirely phonetic, and possibly other versions of it may differ a little. The 'moey and shannon' in the second verse is obviously Moet and Chandon.

Next came the toast of the visitors, which was always placed in the skilled hands of Thomas Trowbridge. He had whiskers all round his face in a fringe, giving him the appearance of a very genial monkey. The visitors usually consisted of the parson, schoolmaster, blacksmith, harness-maker, a keeper or two, and a sailor, retired on pension, who used to measure the men's hoeing each year. Their toast went something like this: 'Chaps, thease be our yearly jollifications, zno, and we can't 'ave 'ee proper wi'out some visitors. And tudn be perlite not to drink their jolly good health. We do know 'em all; fact we do keep the main on 'em, specially passon, and they do know we. And I says we be dom glad to see 'em here thease evenin'. Dom glad we be, and I fears no man when I says that, zno.' And the old fellow would glare round at the company with his whisker fringe all a-bristle with defiance. 'But I bain't no speechifier, 'cepting to say as ow they be truly welcome. Zo I asks 'ee to rise and drink their jolly good health.' This done, the old man would say: 'Now, sit down, all ov 'ee, and I'll sing to 'ee.' There would be loud cries for 'Dick Turpin', and when silence reigned we would hear the following:

DICK TURPIN

1

As I was a-ridin' along on the moor
I seed the lawyer on before
I steps up to 'em, these words I say
Hast thee seed Turpin pass this way?

THE SPACIOUS DAYS

Chorus: *Tibby Hi Ho Turpin Hero*
Tibby Hi Ho Turpin O.

2

No, said the lawyer, an't a seed him this way
Neither do I want to see him this long day
For he robbed my wife all ov ten poun'
A silver snuff box and a new gown

3

O, says Turpin, I'll play cute
I'll put my money down in my boot
O, says the lawyer, 'ee can't have mine
Fer mine's sewn up in the cape behind

4

As I were a gwaine up Bradbury 'ill
I bid the lawyer to stand still
Fer the cape of his cwoat I mus' cut off
Fer me 'oss 'ee want a new saddle cloth

5

I robbed the lawyer of all his store
And bid him to go to law for more
And if my name he is questioned in
He can tell 'em my name is Dick Turpin

6

I am the last of Turpin's gang
And I am sure I shall be hanged
Here's fifty poun' before I die
To gie Jack Ketch fer hanging I.

THE SPACIOUS DAYS

When the applause had subsided the parson would respond for the visitors, and sometimes give us a song himself. Anybody who could sing sang, and a lot of folk who couldn't sing did likewise. On one occasion I sang, and I have a voice like a corncrake. Still, I got through 'Richard of Taunton Dean' with great success.

The gathering broke up at ten o'clock with the singing of 'God Save the King'. Most of the company had to be up betimes next day, for a farm goes on relentlessly in spite of joy or sorrow, sickness or health, good weather or bad. The dairyman and I would be the last to leave, as we were responsible to see that all was safe, and that no match or cigarette end was left glowing, as fire is a real danger to farm buildings in September. When all was safe we would close the barn doors, and stand for a moment looking at the sky in a weather-conscious manner. 'Reg'lar harvest moon, that be. 'Nother blazer to-morrow, zur.' 'Yes, dairyman,' I would say, 'rare weather. Looks as if there'd be time to get in another harvest.'

Around us towered the chunky shadows of the new corn stacks. Bats were flitting overhead. A rat scuttled somewhere in the straw. A three-days-old calf would give a hoarse blare in calling to its mother, who would answer from the home pasture. A flight of wild duck, returning from a moonlight feed on the stubble, would swish overhead in a V-shaped formation. From some nearby field would come the sound of a horse rubbing against the gate. The whole life of the farm seemed subdued by the warm soft-scented dark, but it was there waiting eagerly for to-morrow's dawn.

After wishing the dairyman good night, I would stand

at the drive gate, listening to his footsteps getting fainter and fainter as he plodded down the road to his cottage. An owl would swoop by my face as I went up the drive to the house, and vanish silently in the moonlight. The house looked so solid and permanent against the sky. What a secure, pleasant, spacious business my whole life and surroundings seemed. Rooted so firmly in the soil, surely nothing could ever interrupt or upset its even happy tenor. Farmer's Glory! Farmer's Glory! A swift pat to my retriever, who had slipped a cold nose into my hand, and I would be indoors and soon asleep.

CHAPTER V

Of the agricultural labourer of that epoch I can write only with affection and respect; with affection for his kindliness and courtesy to his neighbours, and with respect for his inviolable adherence to his duty by the soil. Not for wages, nor to please his employer, but because the land was a sacred thing to him, and any neglect was deemed a sin. Possibly the fact that he had no other interest had a good deal to do with this, and another factor was a proper personal pride in his own reputation as a craftsman. I remember once that we finished making a wheat rick on a very windy evening. Granted, when the last sheaf was put in place, it wasn't a very tidy-looking roof, as the wind was nearly strong enough to blow the men off the rick. Still, it was good enough to serve. As often happens, the wind dropped about eight o'clock. My father and I had gone home, but the rick-maker paid some of his mates to stay with him till ten o'clock that evening, pull the roof off and rebuild it more to his liking. 'I wadna gwaine to 'ave no one say as 'ow Bill Toomer built a roof like 'ee wur,' he said to me the next morning when I asked him about it.

A shepherd at lambing times would leave his cottage

and live as a bachelor in the shepherd's house in the field for a month, never going home for one night. And one such I knew, who always shamed me in that at six o'clock each morning I would always find him freshly shaved, whereas I, with every convenience at home, often had a stubble on my chin.

The labourer hated to see any product of the land spoiled, and would go out of his way to stop it, regardless of any consequent effect on his own comfort and well-being. And in extreme age when they were past work, this interest and concern for the welfare of crops and stock in the neighbourhood still persisted. As an instance of this, let me tell you the story of Samuel Goodridge.

Sedgebury Wallop is a Wiltshire village on the banks of the river Avon. It is almost untouched by modern improvements, and, save for one eyesore in the shape of a garage, it presents to the passing motorist the same picturesque serenity as to the passenger in the stage coach of years ago.

Its inhabitants are primarily, nay almost solely, interested in agricultural problems. The farms run from the rolling downs above the village to the water meadows below.

At any season of the year these meadows seem to be the home of abiding peace. In summer one gets peeps of a lush, verdant green between the silvery foliage of poplar and willow. Should one meander gently through them (no other pace is fitting), one finds that they are occupied chiefly by herds of gentle, placid cows and their attendant dairymen, with, here and there, a fisherman or water-keeper.

THE SPACIOUS DAYS

In winter the scene is changed. Both fishermen and cows have migrated to warmer, drier quarters. A rushing torrent takes the place of that tranquil stream of summer days, and practically the only inhabitant is the drowner, undoubtedly a very important person. He it is who manages the irrigation of the water meadows so that the following spring there shall be early grass for the cows. From October to January you will find him, one to every half mile or so of stream, cleaning out ditches, and regulating the water so as to get an even supply to every yard of the meadows. Considerable rivalry obtains between drowner and drowner, for they are craftsmen of high degree, and woe betide any foolish person who does not accord to these experts their due meed of deference and respect.

Now Samuel Goodridge was a drowner. He had lived at Sedgebury Wallop all his life. He started his career on leaving school as a bird starver at two shillings and sixpence weekly, and had graduated through various stages to the position of head labourer at Willow Grove Farm. He was an expert hedger and thatcher, but his chief claim to local fame was that his 'medders' had a bite of grass ten days earlier than any others in the district. In short, he was the king of drowners in that locality.

He relinquished this position very reluctantly, when compelled by the physical infirmities of seventy years of age, to one Bill Yates, a young upstart of but fifty summers. Farmer Wright let him stay on in his cottage rent free, and as his wife was also qualified for the old age pension, and he'd put by a bit of money, he settled down in comparative comfort as a local pundit in drowning matters.

THE SPACIOUS DAYS

Poor Bill Yates had a sorry time. Whatever he did in the 'medders' was wrong, and Sam'll missed no chance of telling him so. He used to toil on his two sticks up to the Red Lion on Saturday nights, and hold forth on the poor state of Willow Grove meadows.

'Poorish show o' grass in Big Maid this spring,' he would declaim to the company in the tap-room. 'Zum volk wun't learn nothin'. Putten stops in main carriages! Thee bist a b——y vooil, Bill Yates. Why dussent get off thee tail water, like I telled 'ee?'

'Oh, gie Bill a rest, Granfer,' someone would say. 'Thee bissent doin' t'maids now. Let 'ee 'ave a show.' 'Show!' Granfer would snort. 'Where's 'is show o' grass?' This sort of thing went on, week after week, and Bill Yates was on the point of giving up his job, when, fortunately for him, the British Army intervened.

A General Danvers bought Willow Grove Farm that spring, and came to live there. He was sixty years old, very precise, spruce in appearance, and as upright as a bolt. He looked on the farm and the village as a hopeless muddle, and he brought the full force of his military training to his rural problems. To do him justice, he recognized the etiquette and customs of the countryside where possible, and also the obligations which devolved on him as the owner of Willow Grove.

Early in his career as a landowner he met Granfer, and told him to carry on in his cottage rent free. Granfer thanked him grudgingly, but in the privacy of his cottage gave voice to his real opinion of the General. ' 'Ee be a meddler, 'ee be. Mark wot I says, 'ee be gwaine t'upset Willer Grove. 'Ee spoke I fair, an' gied I the cottage, but I bain't 'appy in me mind.'

60

THE SPACIOUS DAYS

Granfer was right. The General was a meddler. He purchased tractors and other weird implements. He fenced and cross-fenced. He put in a water supply to the farm, and also to every cottage. Incidentally, it was over this that he and Granfer first came to argument.

The General decided that each cottage must pay the sum of one shilling per annum for water. Granfer objected. 'Dang 'is new-fangled water supply,' he said, 'I got me pump. 'E'll do fer I.' The General went to see him. 'Haw!' he said, twirling his moustache, 'what's all this nonsense about the water charge, Goodridge?' ' 'Tain't nonsense,' said Granfer. 'God Almighty never intended fer a man to pay fer water.' 'Perhaps he didn't,' replied the General, 'but God Almighty never brought it to you in a tap. You'll pay a shilling a year, and don't be such an old fool.' And away he went, twirling his moustache. 'Alright,' mumbled Granfer to himself, 'I pays, but I bain't quite 'appy in me mind. 'E's one o' they volk as can't be telled nothin'.'

In the autumn the General turned his attention to poultry. He bought new and wonderful fowl houses. One afternoon Granfer plodded by the farm, and stopped to rest against the low wall of the yard, from which point he could get a view of his beloved 'medders'. Down in the Alder Plot, a low pasture near the meadows, he saw the General supervising the erection of a number of fowl houses. 'That wun't do,' said Granfer to himself, and opening the gate with difficulty, he toddled slowly down to the scene of operations.

'Be you recknin' to keep vowls down yer in winter time, zur?' he asked the General.

'Of course, Goodridge, these are winter houses.'

61

'Twun't do,' said Granfer. 'I tell 'ee fer why. Thee't be flooded out, two years out o' dree.'

'Nonsense!' said the General. 'I've had new hatch gates put in, and with a little foresight the water can be regulated quite easily.'

'Reggylated be danged,' snorted Granfer. 'I be a-tellin' 'ee. Thee't be flooded. Vive and forty year I looked atter they 'atches, an' when the water comes awver the over-fall at Bickton Mill, t'Alder Plot's swimmin' dree voot deep. You 'ook they 'ouses out o' yer to drier groun'.'

'Well, it isn't your worry, Goodridge. I've gone care-fully into the matter, and decided that the water can be regulated without much difficulty.'

'Gone into the matter, 'ave 'ee? Well! I an't gone into no matters. I don't need to. I do know, an' thee dussent. Thee do wot I tells 'ee.'

'Rubbish,' said the General, and turned away to give some instructions to his foreman, muttering something about silly old fools and anachronisms. Unfortunately Granfer, usually hard of hearing, heard this. It was the last straw.

'Silly owd vooil, be I?' he screeched. 'Nackernism too. Wot be you? I'll tell 'ee. A b——y vooil, thee bist. Fur-riner too. Thee keep hens down yer, an' they'll ha' to turn into ducks, er drown. General thee bist. Well, all I says is thank God we got a Navy. Thee't want 'em, too, to rescue they vowls come winter.

'Look at yer, General,' he went on, 'I gies 'ee a week's notice come Saturday. I bain't gwaine t'ave no landlord, wot calls I a nackernism.' And away the old fellow stumped, leaving the men grinning and the General speechless with annoyance and astonishment.

'What had I better do about his leaving the cottage?' said he to his wife, to whom he related the incident at tea. 'Independent old fool! Do you think he means to go?'

' 'Fraid so, dear,' said his better half, who knew more about rural problems than her husband. 'You see, you've hurt his pride by not taking his advice. He'll never forgive you.'

'Well, what'll I do, dear? These villagers beat me. They're more trouble than an army corps.'

'You've got a lot to learn about them yet, dear. Go and see young Bartram. He's got an empty cottage in the village now that he's milking by machinery. Tell him the whole story. Not that he won't have heard it all in detail by now. He's probably chuckling over it. Still, he'll enjoy hearing it from you. Then ask him to let Granfer the cottage for eighteen pence a week, and make an arrangement for you to pay the difference.'

'I see! Just because I want to keep fowls on my own land, I've got to go to all this bother. Why the——'

'Hush, dear! I know exactly how you feel, but there's no help for it. You can't let Granfer's temper cost him five shillings a week till he dies. What would you say if he tried to tell you how to manage a regiment? But don't arrange for him to pay less than eighteen pence, or he'll smell a rat at once. I was brought up in a Wiltshire village, and I know these people.'

The General, strong man though he was, bowed to the mightier force, and carried out his wife's instructions, thus enabling Granfer to glare triumphantly at him from the garden of his new cottage.

Granfer having won the first round, the second definitely was the General's. The Clerk of the Weather proved to be on the side of the Army, and ordained two dry years.

During this time many things happened. Granfer's wife died, and his daughter Mary and her husband came to live with him. Granfer himself became more and more crippled, and also more crotchety, whilst the General's fowls throve amazingly. Scandals and other topics engaged the interest of the village, and Granfer's flood prophecy was forgotten. But Granfer remembered.

The third summer after the quarrel was an absolute drought. Apart from occasional half-hearted showers, there was no rain from early May to well on in October. The machinery of Granfer's gnarled and twisted body was nearly worn out, and the persistent absence of rain was breaking down his indomitable spirit also. During harvest he took to his bed, and got gradually weaker as the weeks passed by. Doctor Graham told his daughter that the old man could not possibly last much longer, and the general opinion in the village was that Granfer would not see another Christmas.

Then in the last week in October the long drought broke. Warm south-westerly winds lashed the rain incessantly against Granfer's window, day after day. Doctor Graham, who called twice a week to see him, was amazed to find the old fellow decidedly stronger. 'This weather seems to suit you, Granfer,' he said one morning. 'It do,' replied Granfer. 'We've had a dry time, an' now 'tis a levellin' up. We be due fer a mort o' rain, and it be cummen, thanks be. How be the glass, Doctor?'

'Pretty low and still dropping. Looks as if we shall get another gale to-night. Ah, well, I must be on my rounds.

Good morning, Granfer, you're doing fine.' And down the rickety stairs the doctor clattered, while Granfer lay back on his pillows and listened to the rain.

That night the weather broke all records, both for wind and rain. Trees, chiefly elms, were uprooted by the dozen. The mud wall of one cottage in the village collapsed, Farmer Bartram's windmill was blown down, and untold damage was done to thatch and buildings.

Next morning after breakfast Granfer struggled out of bed. Mary, surprised to hear the noise, ran upstairs to find the old chap getting into his clothes. 'Feyther!' she cried. 'Whatever be you a doin'? You get back to bed.'

'Thee find me boots,' ordered Granfer. 'I be gwaine out. 'Tis come at last.'

'What be come?'

'T'water, ye vooil. Cassn't yer thic rumblen? 'Tis awver Bickton Mill, I tell 'ee. 'Elp I downstairs, an' get me boots.'

'But you mustn't, Feyther. Doctor said as 'ow——'

'Dang the girl! I tell 'ee I be gwaine out. Zummut's up. Come on, oot?'

Unwillingly Mary obeyed and Granfer toiled downstairs, and struggled into his boots. 'Now, 'elp I on wie me cwoat, an' gie I me sticks.' Mary did so, and Granfer tottered outside for the first time for eight weeks.

He toiled slowly up the lane with difficulty. When he reached his favourite spot by the farmyard wall, he looked down on the Alder Plot. It was a scene of devastation. Some fowl houses were floating in the flood, and most of the remainder were leaning drunkenly at all angles. Dead and drowning fowls were being swept away, and in the midst of the maelstrom, he spied the General and his gardener trying to rescue some of the hens.

Granfer gazed on the scene for a few moments in silent satisfaction. Presently he saw Bill Yates and another man hurrying towards him.

'Hoy!' he yelled, waving one of his sticks in triumph. 'I telled un, I telled un. Wot about it now? B——y owd vooil! Look at they hens. I knowed I wor right. Ho! Ho! Ho! Nackernism I be. Hoy! General!'

Here he turned and waved his stick at the General. Suddenly he stumbled, and fell to the ground. When Bill Yates and his companion reached him they thought he was dead.

'Poor wold feller,' murmured Bill. ' 'Twer too much fer un. He be gone, zur, I'm thinking.' This last to the General, who by that time had joined them.

They carried him gently down the lane to his cottage, and laid him on the sofa in the front room. As they looked at his withered old figure on the couch, his eyes opened. His gaze wandered vacantly round the room, but lighted up as he recognized the General. ' 'Tis a pity about they hens, General,' he mumbled, 'but I wor right. I be 'appier now in me mind.'

And then he died.

As the desire to be proved right in one's prophecies is one of the strongest forces in human nature, it would seem fairly certain that Granfer died ' 'appy'.

CHAPTER VI

I always thought that the agricultural labourer of those days was very capable with machinery. Of course, he despised any new invention, but when a machine became part of the farm's equipment, he soon learnt how to manage it, and, what was far more important, how to do running repairs with only a very few tools at his disposal. And even twenty-five years ago, farms had a considerable amount of machinery.

Most of this was used only for a short period in each year. For instance, the binders had possibly three weeks' activity, the hay-making machinery about a month, the threshing outfit some twenty to thirty working days, and the corn drill about twelve days' work in the year. But they were all necessary, as to borrow or hire was impossible; all one's neighbours wanted the same machine at the same date. Any breakdown, therefore, was a serious matter, as a day's delay in harvest, hay-making, or sowing, might lose the suitable weather.

So you would find on any farm two or three weatherbeaten peasants, to a townsman's eyes just dull, vacant, and suspicious yokels, who were rarely at a loss in running a machine as complicated even as a binder or thresher.

Whenever I think of them, I am reminded of Pyecroft's remark concerning Hinchcliffe, the engine-room artificer in one of Kipling's stories—*Their Lawful Occasions* I think it is. 'If you hand him a drum of oil, and leave 'im alone, he can coax a stolen bicycle to do typewritin'.' Well, if you gave a farm hand of that period a rusty screw hammer, a bit of wire, an odd bolt or two, and a plentiful supply of 'hoss studs', he could make most farm machinery live up to the reputation its makers gave it. For the benefit of the uninitiated, I should perhaps mention that a 'hoss stud' is a nail used in shoeing horses. It has the merit of being easily pliable, and of having a large head, and, when required, a labourer could be relied upon to produce one or two of these useful articles from some hidden recess in his corduroys. I imagine that they used to purloin a supply of these when the blacksmith visited the farm stables, and probably they picked up others in their travels. . . . They collected any piece of iron or odd bolt in the same way, and stored it away for future use in an emergency. Possibly this was one reason for their stooping gait.

Our threshing machine was driven by a seven horsepower portable steam-engine, which was in the capable hands of one, Thomas Toomer. He died in March 1931, at the ripe age of eighty-four, so that in the days of which I write he was a comparative youth of sixty odd. He ran the engine, and Charlie Bailey, the mower referred to at the beginning of this book, fed the sheaves into the threshing machine.

On threshing days they would leave home much earlier than the other men, to get steam up and everything in readiness for a seven o'clock start in the morning. For this extra work and for covering up the tackle with tarpaulins at night, they received one shilling per day extra. This does not seem an adequate payment in the light of present-day values, but at that date a labourer only received two shillings daily for ordinary time.

They were not supposed to light the fire in the engine if it looked like rain, and rarely was their judgment at fault. Of course, there must always be a great difference in the countryman's attitude to the weather compared with that of the townsman. In town, the weather is only an incident—a pleasing happening or an infernal nuisance. It may affect your decision as to whether you will play golf or go to the pictures; whether you will wear a mackintosh or not; or whether you will have the hood of the car up or down. But it only touches the fringe of life, and the townsman looks upon it merely as an acquaintance of rather doubtful temper. But in the country the weather is the warp in the loom of life. It is your livelihood, your pleasure, your friend, your enemy, and your continuous study. The farming community studies the heavens as the racing man studies the book of form.

I do not suppose there was a moment in the waking life of an agricultural labourer, when he was unconscious

69

of the weather. He noted instinctively when the wind changed, and considered its possible effect on his life and work and upon those of his neighbours. If you met him even on a Sunday, and remarked that it was a nice day, this was no mere formal civility, but the prelude to a discussion of weighty matters. Of course, he 'didn' 'old wi' work on Sunday', but to give advice to the ignorant was another matter. He would gaze critically at the sky, and perhaps warn you to hurry up with your barley sowing as 'the wind be gwaine back agen the sun', and therefore rain was on the way.

Of course, everyone enjoys remarking that it is a fine day, but if, on a wet day, you meet a man who greets you with the remark that it is a nice day, you may rest assured that you have met a true son of the soil, who knows that all sorts of weather are necessary. Such a one I know, a very good friend of seventy-five years. We met the other morning when it was pouring, and he hailed me with: 'Grand rain, this be. Do a power o' good.' I agreed; one does not argue lightly with seventy-five-year-olds. He turned his weather-beaten face to the heavens, and I could see the raindrops splash on his forehead, and run down the side of his nose into a forest of whiskers as into a sponge. 'Ah!' he said with satisfaction. 'Wind's droppin' back. It be cummen, thanks be.' And after a few moments conversation he stumped away quite convinced that God was in his heaven, and all was right with the world.

But I am getting away from threshing. It was hard work. There was a balanced team of men to run the outfit. Two men pitched the sheaves on to the top of the thresher, where one man cut the strings and another fed the grain steadily and smoothly into the machine. Two men made the straw rick, one took off the chaff, and one the grain, whilst Tom, the driver, was in charge of the machinery. Any one man slackening speed would slow up the whole business. No extra money was paid to the regular farm men for threshing, but beer was allowed on threshing days, one quart per man. Personally I did not like threshing as the dust made me sneeze continuously all day, but I have done every job in connection with it, even to minding the chaff or dust, as it was termed locally. That was a dirty job. You stood in a narrow passage-way between the machine and the corn rick, and worked in a dust cloud all day. The man who usually did it had whiskers growing out of his nose, and these would be festooned with cobwebs in the first hour.

Tom and Charlie must have known much more about machinery than I gave them credit for in my unthinking youth, as I cannot ever remember a serious breakdown, and their reliability was extraordinary. On the evening before a threshing day you just said: 'Oh, Tom, we shall be "sheening" to-morrow,' and, if fine, when you arrived at the ricks at 7 a.m. on a January morning, say, you would find Tom standing in the glow of the fire with wisps of steam playing around his head. And the old boy had left his home at 5 a.m., and walked perhaps a mile and a half to the ricks!

The first arrivals would greet him with: 'Marnin', Tom. You do know zummat, you do. I 'low Tom do meet a young 'ooman, zno, up yer in marnin'.' And old Tom would grin a sheepish grin, as if he had really been guilty of this suggested early morning dalliance with the fair. I think he had been involved in some moral pecca-

71

dillo in his riotous youth some forty-five years before, and they would never allow him to forget it.

The foreman's greeting was of a much more dignified nature. 'Marnin', Mr. Toomer. Weather gwaine to wear?' Perhaps it would be drizzling, but Tom would reply: 'Ay! Thee cannst get 'er stripped. Wind's gwaine round wie the sun. This yer flit doan't mean nothin'.'

'Now then, chaps. Let's 'ave you,' the foreman would say. Some men would strip the thatch from the rick, whilst others would pile it out of the way of the machinery. Charlie would mount to his feeder's place on top of the thresher, and his mate would sharpen his knife on a whetstone. When all was ready, Tom would shout: 'All right?' 'Ay, Tom. Let her goo.' Tom would push the throttle lever a little, the driving belt would tighten, clap once or twice, and in a moment or two the rich hum of the thresher would fill the countryside, and dominate all of us till lunch time at nine-thirty.

Ricks which were not threshed until after Christmas usually contained some rats, and this added a spice of sport to the finish of the day's work, as most of them remained hidden until the bottom layer of sheaves was moved. They were killed ruthlessly; a pitchfork is an efficient weapon for this purpose. Some would leave the rick and perhaps get into holes in a bank or hedge near-by. These were bolted by pouring boiling water from the engine into the higher holes, thus driving them out of the burrow in a semi-scalded condition. This may sound cruel, but rats were rats, and as such, enemies to farming, so they received no quarter.

At some time in the forenoon Tommy and the trap would arrive bringing the master mind. My father would inspect the corn by dipping his hand into a full sack, a method of judging whether the corn was dry; you cannot push your hand into a damp sack of wheat. A handful would then be smelled vigorously. I can see them at it now, with grains of corn adhering to my father's moustache and the foreman's whiskers. Before he left father would study the portion of the corn rick still unthreshed, with a shrewd, calculating eye, and say to the foreman: 'Well, you'll be able to finish all right in nice time?'

The foreman would never give a definite answer. He knew the variable quality of the weather, of the machinery, and of countless other influences which might hinder the work. 'Well, we mid, zur, if we do 'ave luck.' Nothing more definite or hopeful was ever forthcoming, but, barring accidents, the rick was usually finished in the day. Most of the men went home on the top of the last load of grain at the finish of the day, but Tom and Charlie stayed behind to cover up and to see that all was safe and ready for next day. Tom always walked home, using a piece of dry wood as a walking stick. It might be of any shape, like the one used by Sir Harry Lauder, only more so, but he walked from his home to his work without any stick, thus taking home a piece of dry firing each day.

The only definite holidays for the men during the year were Christmas Day, Good Friday, and the local fair day. On the first two the essential jobs in connection with the livestock had to be done, the men not regularly in charge of any stock giving a helping hand to the stockmen to enable them to finish early. On Good Fridays they planted their own potatoes in land set aside for this purpose each year. Ten rods were allowed for every married man, and five rods to bachelors and boys. The milkers' potatoes were planted by the others, as the milkers could never get off.

On fair days the men drove the sale sheep to the fair in the early morning, and were then finished for the day. This fair was a business one in the morning, and a pleasure fair in the evening. In addition to sheep selling it was the recognized hiring fair. Men in search of a new situation wore the badge of their calling in their hats. A carter wore a plait of whipcord, a shepherd a tuft of wool, and cowmen sported some hair from a cow's tail.

Men were hired from Michaelmas, October 11th, for the year, usually by verbal agreement, the essential features of this being noted down in the farmer's pocket-book. All sorts of things came under review during the discussion of these agreements. You might agree with a man subject to a favourable character from his present employer, to find on inquiry that although he was all right, his wife was of a quarrelsome disposition. This might be a hopeless drawback. A farmer doesn't just employ a man, and remain in ignorance of his life during non-working hours. He has to live with him, and these domestic differences can upset the whole farm. In many cases the cottage available might be one of a pair having a common front door opening on to a passage between the two houses. Some blocks of cottages had only one copper and washhouse, so that washing days must occur in rotation. Still, if it happened that you had a single cottage available a bad-tempered wife did not matter.

I can remember one carter agreeing with my father

subject to the cottage chimney not smoking. 'I've a lived in a smoky 'ouse fer seven year, zur, and my missus do say as 'ow she bain't goin' to another less she do know the fire do draw nice and suent. You do know 'ow 'tis wi' wimmenvolk.' He and his wife journeyed over a few days afterwards to inspect the chimney, which, I remember, proved satisfactory, and the man worked for my father for several years.

The pleasure fair in the evening was a whirl of round-abouts, swinging boats, coconut shies, shooting galleries and sideshows and cheapjacks of all kinds, the whole place being lit up in the evening with reeking naphtha flares. Here rural youth made high holiday.

In my childhood days, I can remember being lifted on to a bench, and, kneeling there, shooting with an airgun at a bull's eye on a large box some two yards away. When you hit the mark, the lid of the box flew open, and a large stuffed monkey on a piece of elastic jumped out.

Then there was the 'Bombardment of Alexandria' in a large tent. This was a glorified magic lantern show, the forerunner of moving pictures. I have never enjoyed any London theatre more than I did that crude entertainment in my boyhood.

There were 'Try your strength' towers, where you drove a weight with a large mallet up the tower where it rang a bell if you got it to the top. Farmers and labourers vied with each other at this trial of strength. One of our men could do this one-handed with ease, and used to coach the inefficient, much as a golf professional in later years supervised my beginner's efforts. The words used by each teacher are different, but the meaning is the same. 'Doan't 'ee goo at un zo ravish.

You wants to take it easy and suent, zno, but you wants to ketch un jist right. Like this.'

Later on in the evening the crowd would be packed shoulder to shoulder, there would be confetti in profusion, and some horse-play amongst the younger labourers. But even in the midst of all this riot you would find the older men talking about their work. Wedged by the side of a shooting gallery, one shepherd would say to another: 'Our swedes be good t'year, and we'm got plenty o' hay, so we shall do all right thease winter, I 'low. How be you fitted fer grub thease season?' 'Oh, 'low we shall manage. Wonder where my missus be? I do want to get whoam. We be shiftin' flock to-morrow.'

Women and boys might enjoy this yearly revel, but the older men's interest was in the land and its needs.

A General Election was another occasion which obtained a half-day's holiday for the men. They voted at a village schoolroom about three miles away, and used to journey to the poll in a two-horse wagon. My father would not allow any of the political parties to take them, and insisted on the team of horses being decorated with red ribbons, the Liberal colours, on one side, and blue ones, the Conservative hue, on the other. 'And we do get booed gwaine up and comin' back,' one of the men once said to me.

Generally speaking, the men were Liberals and the farmers Conservatives. I think that agricultural labourers will be Free Traders for a good many generations to come, for even to-day the older men can remember the poverty and shortage of food during the last protectionist epoch. Old Dick Turpin, who came from a family of shepherds, told me once that in his childhood he was fed on barley bannocks, and that a small piece of bacon once a week was the only meat which came into his home apart from a poached rabbit or a dead sheep. His father was allowed so much coarse barley meal each month to feed the sheep dogs. His mother used to sift this through a piece of muslin, use the fine sample for the children, and the coarse for the dogs. The cry 'Your food will cost you more' will influence rural elections until that memory has died away.

My father was a Liberal, chiefly from conviction, but partly, I always thought, from the joy he got in being different from his neighbours. His landlord's family was of the true blue Tory type, but their relations with him were of the most friendly and courteous nature. Full-blooded opposition they enjoyed and respected, but the weight of their influence and power at election time was felt by a good many people in the neighbourhood. Generally speaking, apart from the labourers, the politics of the ruling landed house were the politics of everybody else. Sometimes it happened that the politics of the ruling house changed with the death of an old landlord, the inheritance of the estate going to a man of different political views. I know one district where the new man changed from Liberal to Conservative and the whole district went to bed one night strong Liberals, to wake up next morning and discover that they were Conservatives, or had better be.

The letting of farms in those days was a personal, friendly thing, not solely a business proposition as it must be to-day. All sorts of considerations came into the choice of a tenant in addition to his farming capabilities. Some estates insisted on a prospective tenant being a Conservative and a Churchman, and of bringing definite evidence as to this. A guaranteed sufficiency of capital was required; one couldn't have one's tenants going wrong financially, as people would think you were charging them too much in rent. Non-interference and help with the landlord's sport was another deciding factor. With a hunting landlord an absence of wire was more important than good farming. With a shooting landlord growing a plentiful supply of root crops as cover for his partridges, and training one's men to report all nests of eggs to the keeper of the beat, and to abstain from poaching, was another desirable qualification.

And usually the landlord played fair. If a tenant did all he could to further his landlord's sport, he found it easy to get repairs and improvements, and, in addition, he got a goodly share of his landlord's sport. Apart from a few large house parties our landlord made it a practice to invite the tenants, whose farms were to be shot over, to join the party on the particular day. Looking back on those days, it seems to me that the shooting was more important than the farming. But each was necessary to the other. It was a case of live and let live, and of working in harmony together.

But for a large farmer to be an avowed Liberal was a black sin generally, although my father got away with it successfully. A portion of the old feudal system still remained, and for a tenant farmer to do anything against his landlord's wish was unthinkable in rural circles. At one election I can remember a neighbouring farmer in the heat of political argument saying to my father, 'I can't make you out, Mr. Blanchard. You hunt, you shoot, you're a sportsman, and yet you vote Liberal, and rent a farm from the Duke of ——.'

'Well, what's my voting got to do with my landlord?'

'Do? Why, you ought to support your landlord. Where would —— be', naming my father's village, 'without the Duke of ——?'

'Why, where 'tis now,' my father replied.

'No, 'twouldn't,' shouted the other. ' 'Tud be in hell, where all you Liberals ought to be.'

This was the time of the Lloyd George Budget elections when feeling ran high, and there were sometimes some ugly scenes at the declaration of the poll—usually, I must confess, when the Liberal got in.

In those days any land, even small pieces, which came into the market, was usually purchased by the owners of large estates, in order to prevent interference with their shooting, and I think that one of the reasons for my father's Liberal opinions was that he had a hankering to farm his own land, thus being beholden to nobody, although I have heard him say that to rent under a good Tory landlord was the next best thing. He was always inspecting farms which came into the market, but as these were few and far between in those days, and any farm he might purchase had to compare favourably with his present renting conditions, he never found one to suit him.

My mother came into this argument, too. Our farm was situated only three miles from a large market town, and had a large and convenient house. These were the days when the motor car was in its infancy, and my mother refused to be stranded on some farm perhaps twelve miles from a town, no matter how cheap it might be. Once my father found a real bargain so situated, and tried to persuade my mother to let him buy it. 'It's dirt cheap, Mam. If I farm it well for four years it'll be worth double the money.'

'It's out of the world,' said mother. 'Besides, we don't want it. We're very comfortable here.'

He tried to bribe her to go there with the offer of a motor car and a chauffeur to take her to town when she liked. That proved unsuccessful for my mother was too well versed in farming enconomics to be caught like that. 'Motor car and chauffeur,' she snorted. 'And when I wanted him he'd be at the other end of the farm doing some job for you. Catch you keeping a man lolling about the buildings all day, in case I wanted him.'

As a last resort he tried the high-handed method on the lines that women were all very well in their way, but could not possibly understand real business. 'I shall buy it, Mam,' he said one day. 'It's too good a chance to be missed.'

'Well, you buy it, if you want to, but I shan't go there with you, and that's that.'

I have always wished that he had said that he would go there alone, as I should have liked to see how my mother would have handled that situation; but he knew better, and bowed to the mightier force. Subsequent events proved that he was right in every detail in his estimate of the bargain, and I know that he always regretted not buying that farm. But that was very definitely that.

Another time he became enamoured of a farm some forty miles away, in another county, and as my mother was a great churchwoman, he thought he would get her consent to buy it, as the freehold of this particular farm carried with it the gift of the living in the village. 'Think, Mam. You'll be able to choose your own parson, and you can see the church from the dining-room windows.'

'I've got enough to see to without bothering with parsons,' my mother replied, 'and I'm too old to go into a new district away from all our friends. You're like a troubled sea, never still.'

And again, that was definitely that.

He never did buy a farm, but I have often wondered what he would have done had he lived to see the major portion of the farms in his district up for sale.

In the foregoing part of this book I have tried to give a true impression of the prosperous and secure lives of tenant farmers some twenty-five years ago, and before leaving this period, I would again point out that although they had plenty of sport and pleasure, it was all connected with their calling. It was possible in the winter for a large farmer (and most farms in the sheep districts were of necessity of at least four hundred acres) to hunt two days a week, shoot two days, go to market one day, and possibly to earmark the remaining weekday as pay day, which would necessitate driving to town to cash a cheque.

Mind you, on all these days he would make the round of his farm in the early morning, and during the pursuit of these various sports, he would be riding or walking over farms in the district, or talking to people who had an intimate connection with agricultural matters.

It was much the same in the summer time, when tennis parties were happening nearly every day. One played tennis hard; there was always joy in competing with one's neighbours. But in between sets one talked farming, and often slipped away with one's host to inspect his mangolds or his stock.

Rarely did a farmer sleep away from his farm. 'An hour in the morning's worth two at night,' was the ruling motto. In the morning you set your men, of whom you had a plenteous sufficiency, to suitable jobs in accordance with weather conditions, and you could get off for the remainder of the day pretty frequently. A farmer's life may have been ruled by the seasons of the year, but it was not ruled by the clock like a city dweller's. There were no hard and fast hours. At certain seasons one worked eighteen hours a day, and at others one worked according to one's own estimate of the necessity.

For a townsman to have enjoyed the pleasures which came almost automatically to a farmer, would have been very expensive, but in the farmer's case they entailed scarcely any cash outlay. For instance, a horse to ride and drive with a man to look after it was part of the general expenses of the farm. The groom would do a lot of essential farm work in addition to tending the horses. which themselves were used only partially for pleasure.

A farmer's subscription to the local hunt was the permission to ride over his farm. Shooting entailed only the cash outlay for cartridges, and judicious half-crowns as tips to keepers and beaters. These last were a sound business investment from a farming point of view. When properly given and received, the recipients never omitted to do all sorts of helping things for the farmer. Gates left open by picnickers would be closed. A keeper would walk two miles maybe to give information about any stock requiring attention at the top part of the farm on his beat. The village poacher, who might be making a nocturnal raid on your rabbits, would spare no pains to put anything right for you on his travels, a horse or cow in a ditch or a sheep on its back. This sort of thing was a feature of rural life, which could not have been carried on satisfactorily without it.

The large tenant farmer's social position was peculiar. Definitely he was not 'County'. There was a distinct line drawn between the owners of land, and those who rented it. But the 'County' met him as an equal over rural sport. In his own opinion the farmer was very superior to anyone in trade, I mean, the retail trade necessitating the trader keeping a shop, and, horror of horrors, serving behind its counter. To clean out a manure yard was a gentlemanly occupation by comparison. This is dying hard. You will still find in country districts tennis clubs and other societies which refuse membership to anyone who may be discovered during the week serving behind a shop counter, whilst farmers are accepted gladly.

Consequently the ruling house of the district gave three balls every season: one for their own friends, solely a 'County' affair, a tenants' ball, and a servants' ball, which last embraced the despised shopkeeper.

There were, of course, some unfortunate folk who did not fit into either of these balls, and some of them had the privilege or suffered the indignity of being invited to the last two.

Truly, the drawing of the various 'boundary' lines was a real and lively problem in rural circles.

CHAPTER VII

I have been told that it is useless for a farmer to try to write a book, as he is almost sure to leave out the things which are most interesting to the general reader, deeming things which are commonplace to himself as unworthy of mention. Another danger is that his book will develop into a sort of textbook on agriculture. I do not think that I can here be accused of the latter charge, and I will in this chapter try to recall some odds and ends of happenings which may justify insertion.

The song of the Fourth Wiltshire Regiment is still 'The vly be on the turmut', which extols the delights of hoeing as an occupation. I am afraid I do not agree at all. I have done some hoeing, and it cured me of any desire to sing about it. Pursuing his policy of making me do every job on the farm at some time or other, my father suggested one summer that I should hoe a rudge of swedes and kale. Not, mark you, that I should hoe it in the ordinary working hours. Oh, no! This was to be a job for after tea, and I was to be paid at the same rate per acre as the men. It needs no effort of memory to recall that price. It was six shillings and sixpence per acre for flat hoeing and seconding, and seven shillings

84

per acre for singling, or as the men put it, a pound for three times.

The usual practice was to take in to forty-two drills, which meant about a chain wide, and as this particular field was twelve chains across, my piece came to just over the acre.

As the men said, I was too long in the back (I am well over six feet) and I suffered accordingly. First we flat-hoed between the rows, cutting as close as possible to them. The prevailing weed was called stoneweed. I do not know whether that is its correct name, but it will serve. It appeared to be constructed of chilled steel wire. Evening after evening in July I beat and hacked at it, feeling very hurt in my pride that men of sixty were hoeing close by at double my speed, and making a better job. I had always found that once seasoned to a job, I could hold my own with any of the men, but at hoeing I confess my inferiority. I was then, and am now, a bad hoer.

Flat hoeing finished, we turned round, and singled the plants about fifteen inches apart. This looks a horrible mess just after you've done it, but next day all the plants you have left have their heads up to the sun, and look splendid, whilst the other nine-tenths you have cut out soon wither away. What a sigh of relief I gave on the evening that I finished singling my piece. I had just learnt to throw a fly, and wanted to take my rod up to the meadows in the evenings as often as possible before harvest. How light and flimsy it seemed, too, next evening, compared to the hoe.

But it wasn't finished with my hoeing. About ten days after singling, the order was given that the field must be

seconded. This meant hoeing all the ground, both between the rows and between the plants, and singling out
any double plants which had been left the first time. To
this day I stick to it that the field did not need this extra
hoeing. It was as clear of weeds as a front flower bed.
But the harvest wasn't quite ready, and I am sure that
my father and the foreman hatched this plot for my
especial benefit. Anyway, we seconded the field, whether
it needed it or not, and I have fought shy of hoeing ever
since. Also, I confess that in my future dealings with
hoers I have always paid on the generous side.

Some people say that successful men are lucky. It
often seems so, but I think that they have a special
genius, a flair if you like, for doing the right thing, which
cannot be defined exactly. I know that my father had
this gift in a marked degree. It was our general custom to
break in a cart colt or two each season. I had watched
this proceeding several times as a boy, and it had always
happened all right. Shortly after I left school, we took a
colt out one morning to break it. When I say we, there
were the foreman, and three carters in charge of the
business; I was there for education, and the 'organizer'
was absent.

In this particular case it didn't happen according to
plan. With much 'woaing' and many 'stand still, oots',
the colt was hitched to a plough alongside an old steady
mare. The carter took hold of the plough handles and
away they started, or should have started, with a few
plunges from the youngster. But that youngster refused
to budge. The old mare went away on the word of com-
mand, but the colt stood fast. They coaxed him with
'Now then, little feller' and other endearments, to no

purpose. They whacked him, they swore at him, they made horrible, sudden, weird noises and catcalls in his rear, but there he stood, hunched in sullen immobility. The old mare would look round at him with an inquiring eye as if to say: 'Come on, I can't stay here all day.' Gradually the men lost their tempers and became more cruel in their methods of persuasion, but it was no good. 'Scoatin' little bastard,' said one of the carters. ' 'Ee do beat all. Blast ye. Now then, coom up, Vi'let, altogether.' Vi'let came up nobly, but the colt was not included in the altogether. He evidently disapproved of team work. He was as the cat who walked by himself, save that he didn't walk, but stood by himself. Verily, all places were alike to him, for he showed no desire to go anywhere. Finally, the breaking was abandoned. 'I doan't like gie-in' up,' said the foreman, 'but I bain't goin' to have 'ee beat no more. Dang un. You can do anything wi' 'em if only they'll goo, but when they wunt goo at all, you be done like. The Guvnor'll create about this, though.' The horses were unhitched, and, just as the disconsolate procession was leaving the field, my father appeared. 'What are you coming away for?' he asked the carter who was leading the colt. ' 'Ee wunt goo, zur, nowhow.'

'Won't go? Then he must be made to. Come on, back you go.'

And back we all squelched to the scene of our failure. My father's manner seemed to infuse a feeling of brisk-ness into the proceedings. During the preliminary hitch-ing he gave everybody the impression that this was a trivial business which would soon be over. As his son I knew jolly well that I would never dare to stand still if he wanted me to go anywhere, and I began to be a bit sorry for the colt. I felt that he couldn't possibly realize what he was now up against. When all was ready my father gave one last vicious instruction to the carter. 'Now then, Tom, when he starts let un go. If you holler woa, I'll sack you. I'll tell 'ee when to stop. Now then.'

I think everybody present hoped that the colt would refuse to budge. I know I did, as I wanted to find out the way to overcome this difficulty, but I can remember feeling quite sure that my father would have a certain cure for it.

'Noo then, Vi'let, coom up,' said Tom. As before, Vi'let came up in great style, but her companion failed again. But this time he went one better than just standing still. He went backwards a bit and sat down heavily.

Now a cart colt weighs nearly a ton, so if you can imagine a very fat man about twenty stone sitting down in a chair with a bump upon the point of a long sharp tin-tack, you will get some idea of what happened. There are lots of spiky things on an iron plough, and the colt sat down fairly on one of them (the drail pin is its correct name) and ran the point some three inches into his ham.

He didn't sit for long. No, by golly, he was up and away with Vi'let doing her best to keep up, and Tom hanging on to the handles for dear life. The other men ran alongside over the rough ground in a stiff, scrambling, awkward gait—one's knees do not bend easily for running at sixty or thereabouts—whilst my father gallumphed along behind in like manner, giving tongue with hunting calls to cheer on the horses. 'Keep 'em going, Tom,' he puffed. 'Don't let 'un stop on his own. He must go till we want him to stop.' And they kept him at it until he was a lather all over, and afterwards he was very little trouble. He was christened 'Squatter', and carried the mark of his squat all his life, as although he was a bright bay in colour, a small patch of white hairs grew over the wound in his ham.

Having demonstrated how easy it was to make a stubborn colt go, my father lectured the men on the imbecility of hitching off a colt before conquering it, and set off for another part of the farm to wake up somebody else. I stayed behind. The men's comments were much the same as my own would have been. 'Guvnor, 'ee do take all the credit for thic young vooil runnin' thic pint in his backside. Why hadn er bin yet at the beginnin' and showed us how to do it? I 'low that ud a beat un. But there, things do allus goo right fer 'ee somehow.'

Things did go right for him, somehow, even his mistakes. He was great on checking everything one wrote. 'Check and double check', he would say to me again and again. It was relic of some far-off business training in his youth. I remember once making a frightful mistake in an order, owing to omitting this double check business. I had written:

2 gross Cooper's Dip
6 —— White Oils

meaning of course six bottles only of the latter commodity. When the consignment arrived I got it pretty hot. You know the sort of thing. 'Can't I leave a simple thing like that to you? Perhaps you'll listen to me next time I tell you to check what you have written.'

However, he got the firm to take back most of the six gross of oils, and there wasn't much harm done. And then, shortly afterwards he made a slip in ordering himself, and I felt that the Lord had delivered the enemy into my hands at last. He was away on holiday when I got a notice from the station that forty tons of linseed cake had arrived. I knew that only four tons was expected, but I had no alternative but to get the lot hauled home. I wrote to my father fully about the matter, and took a great joy in composing the letter. A few days afterwards he returned home, and next day after breakfast, when I went up to his bedside to give him my usual report, he said sardonically: 'I suppose you're rather pleased that I made a mistake over that cake?' My face betrayed my satisfaction. 'H'm,' he went on, slapping the newspaper. 'We all make mistakes some time or other, but when I make 'em they're worth while. Look, cake's up twelve and sixpence a ton. Tell Bill to have the trap outside in a quarter of an hour.' I did so, and we went into the market town where he got seven and sixpence a ton profit on thirty tons of the cake from a merchant, who hauled it in to his store the next day. Pure luck, as my father admitted to me some years afterwards, but very impressive at the time to my inexperienced youth.

Of course, not all his mistakes were profitable. Against the foreman's pleading he carried some barley one harvest, before the butts of the sheaves, which were full of clover, were sufficiently withered to be safely ricked. But we had come to the last field, the weather was perfect, and there was the whole harvest team in working order with nothing to do. So we put eighteen acres of barley into two ricks each nine yards long by four and a half yards wide. There was no joy in the doing of it, either. The men knew it was wrong, and were sullen and sour. My father knew it was wrong, but wouldn't go back on his first decision. Two days afterwards vapour was coming from the roofs of the ricks as from a chimney. A long iron rod was thrust into the heart of each rick, and left there for two hours. When we pulled them out the ends were too hot to hold. 'Nobody's fault but mine,' said my father. 'Well, we'll have to pull 'em apart.'

We did so, and the scent of the heated barley in them comes back to me as I write. It was like a malt house. We made those two ricks into eleven little ones, placed carefully at the back of the buildings out of sight from passers in the road. The two original ricks were very hot, and the heat made our feet very tired and sore, as we stood on them and worked all that day. Later on when the little ricks were threshed the barley was ginger brown in colour. It was valueless for malting purposes, and was crushed and fed to our own stock. One didn't blazon one's mistakes abroad to one's neighbours by exposing such a heated sample of corn in the market. But rural communities had even in those days a means of communication which beat Marconi's, and my father got a good deal of chipping about this business from his neighbours.

Still, it was his mistake, and his pocket was the only one to suffer. And it could stand it. There was a margin of profit in those days with which to take chances, and I admire my father more and more as I get older. He was a good winner and a good loser, and he was always scrupulously fair. Moreover, he was always boss of his job. Not only on the farming side, but he interfered in the men's private lives when occasion required. The foreman reported once that one of the men used to get too much to drink of a Saturday night, and go home and beat his wife. 'Didn' ought to be, zur.' At that time my father was rather badly crippled with rheumatism, and the man in question was a husky chap about fifty. We drove along to his cottage that evening, and I sat outside in the trap whilst my father hobbled up the garden path, and knocked at the door. For any normal purpose he would have stayed in the trap and had the man come out to him, but this was a man-to-man thing. The man came to the door, and my father bullyragged him unmercifully. 'I'd sack 'ee this minute if it wasn't for your wife,' he wound up. 'Now mind, one more word about this sort of thing, and I'll come down and give you a damn good hiding in front of your wife. Pah, look at your garden, full o' weeds. But there, a coward who'd beat a woman is never much good at anything. Now mind, if I hear you've been down to pub till I give you permission, I'll knock your head off, whether I'm up for assault or not.' And such was the awe in which he was held by the men that the wife-beater not only stayed away from the pub, but turned teetotal soon afterwards, and became a model husband. I used to wonder what would have happened if my father had had to carry out his threats of a hiding, as he would have stood a poor chance physically.

I can see now that it was inevitable that we should come to serious argument as time went on. I was most certainly an insufferable young pup in many ways, as, I think, are most of us at eighteen or thereabouts. Anyway, some two years after I left school we came to the parting of the ways. My idea was that I had become a sort of errand boy between my father and the foreman, and that this was hardly good enough for a man of my qualifications. My father's idea seemed to be that he was blessed with a half-wit for a son, at least that was the thing he was always most careful to infer at all times in order to keep me in my proper place. Finally the thing flared up into open rebellion. I said that I was tired of being an errand boy, and wanted a job on my own with some responsibility. Privately, I thought that either my father or the foreman, preferably both, should retire, and leave me to show them a thing or two. My father said, more or less like Bairnsfather of later date, that if I knew of a better job, why not take it? Youth's pride being mortally injured, I said that I damn well would, and was rebuked for swearing in addition to my other crimes.

Chance so willed it, that before I had time to cool off, I found in the local paper an advertisement for a young man to go to Canada, and I wrote for particulars. A day or so later I got a letter from a Mr. Hartley. He was a solicitor in the north of the country, and was trying to get a lad to go out to his son in North-west Manitoba. After some correspondence between him and my father, who privately thought that each of us had bluffed a bit too far, but that anyway it would do me a power of good, the whole business was fixed up, and on 5th September 1911, I stood on the deck of the S.S. *Corsican* of the Allen line, watching my father's figure getting smaller and smaller as the tugs towed the liner down the Mersey.

PART II
A CANADIAN INTERLUDE

CHAPTER VIII

Any details of the voyage seem to be non-existent. I can remember being seasick for one awful twenty-four hours, but I soon recovered and was able to do justice to the excellent fare provided. I had five pounds in my pockets and twenty-five sovereigns in a leather belt round my waist next my skin. Although this was extremely uncomfortable, it made me feel a bit like a pioneer going out to savage parts, and I had visions of great doings in the wild and woolly West, in which I should be the central figure. I can remember feeling extraordinarily alone. For the first time in my life I was in a world where no one knew who I was, and seemingly cared less. England and Partridge Farm seemed very far away.

My cabin mate was a mining engineer named Curtis. He proved a good Samaritan in many ways. He showed me round the boat, fixed up our places together in the dining saloon, and gave me lots of good advice. 'Shouldn't drink much if I were you, son, and don't play cards for money. I shall do both, I guess, but I'm past praying for.' Again, as we passed two pretty girls: 'If you must spoon in the evening, choose chairs that haven't got a light over 'em, or you'll feel an awful fool when they're switched on.' As I was only eighteen this last piece of advice was wasted. I was much too scared to engage in light dalliance with any of the damsels. Besides, they had no lack of cavaliers.

We landed at St. John, and I got through the customs farce very easily. I had booked right through from Liverpool to my destination—Barloe, Manitoba, Canada. The purser of the *Corsican* had returned to me a long strip of paper, which, he informed me, was my railway ticket. The train journey seemed endless. Not knowing anything about Canadian travel, I had booked at the cheap Colonist rate, and the train was full of men of all nationalities, going West for the harvest. They fought and swore in many languages.

The sleeping accommodation was crude. I slept in a hard rack which was pulled down at night as a sleeping bunk. Sometimes I would be awakened by the train shunting; not ordinary shunting as practised in England, but just as if the engine-driver was in a bad temper, and was trying to shake off a carriage or two. It was usually in the middle of the night that the officials examined our tickets. Each time this happened my ticket got shorter and shorter, as pieces were torn off, and I arrived at Winnipeg with a piece only about an inch long.

By day, the journey was uninteresting. After all, I was

a farmer. But where were the golden wheat fields as depicted in the shipping posters? All I could see from the train was scrubby trees and rocks. The land seemed to be almost solid rock. Every now and then one of the fence posts at the side of the track was propped up with boulders as evidently no hole could be made for it. Grain land, I gathered from my fellow travellers, did not start till west of Winnipeg.

In places the track skirted the great lakes, and the train ran along a ledge cut in the side of the mountain. In one place I saw, lying in the lake below, a train which had evidently rolled into the lake at some time or other. Some years afterwards when I came back it was still there. After three days of this I arrived at Winnipeg about 7 a.m., and found out at the inquiry office that a train started for Barloe at ten. I had a wash and a good breakfast, both of which I needed, and filled in the remainder of the time watching my surroundings.

Winnipeg Station, at any season of the year, presents a varied and interesting throng to be studied. The men were of all nationalities, but they nearly all had one thing in common; they were clean-shaven. There were townsmen in high-shouldered suits and boots with funny bumps on the toecaps, leathery-faced miners, lumberjacks, and teamsters; excited French Canadians come up from the East for the harvest, North-West Mounted policemen, one or two cowboys in high-heeled boots and Stetson hats, and one Indian in a blanket and moccasins, who surveyed the bustling crowd with haughty indifference. There were various types of peasants from central Europe with their families. Women were few, and these were chiefly Galician peasants' wives, who sat amongst their brood waiting patiently, oh, so patiently, for someone to tell them when to continue this interminable journey from Austria. Ten o'clock came along very quickly, and after walking across several tracks, I climbed up the steps into the train.

Barloe is about one hundred and fifty miles west of Winnipeg, and we stopped at Portage la Prairie, which is about half-way, for lunch. 'Dinner, son,' said the conductor. 'Come on, we've got twenty minutes.' We had a good meal at an hotel, during which the conductor found out that I was going to a farm near Barloe. 'You'll be in time for threshing,' he said. 'Most all the crop's cut, but they ain't threshed much.' The engine bell recalled us, and I continued my study of the countryside.

Barloe was reached about 3.30 p.m., and I was the only passenger to alight. 'One green Englishman, and Jack Smither's pair o' boots, is all for you 'sides the mail,' I heard my lunch companion say to the station agent. The pair of boots, I discovered later, was in reality two bottles of White Horse whisky in a boot box, consigned as boots to overcome the local option regulations, as Barloe was dry.

The station agent grinned cheerfully at me as the train pulled out. 'Where you makin' for, son?' Everybody called me son, so I suppose my youth and innocence was pretty apparent. 'Mr. George Hartley's farm,' I answered.

'George ain't in to-day. He's still busy cutting. Still I guess there'll be someone in for mail who'll give you a lift. Follow old Mac over to the store, and ask him to fix you up.' He pointed to the postmaster, who was walking across the track with the mail bag. I did so, waited till the mail was sorted, and then inquired of the postmaster as to the best way to Mr. Hartley's farm.

'Hartley's, laddie? Ay, George told me ye was coming. There's Henderson now, he'll do yer business fine and dandy. Say, Pop!' At this a man at the other end of the store came up to the mail counter. 'This young mon's goin' to George Hartley. 'Tis a wee bit out o' yer trail, but if ye're no busy, Pop?'

'Sure thing, Mac. I can go that way. I'll pull out right now.'

'There y'are now, all fixed up,' said Mac. 'Bide though, ye can take George's mail.'

Having pocketed some letters and a newspaper, and thanked the postmaster, I followed Henderson out of the store to a waiting two-horse buggy, and we set out.

As we drove along the trail between stooks of grain, chiefly oats, Henderson inquired my name and I let out in conversion that I was a farmer's son. 'Then you'll be some good to George this Fall, p'raps. Gordon, that's the fellow who's been batching with him, is off to the Peg next week. Say, do you farmers in the Old Country ever do any work?' I said that I thought so.

'H'm, think you could drive a stook team?'

'What does that mean exactly?'

'Wagon with a sheaf rack on. Get a load of sheaves, and pitch 'em into the separator.'

'I think I could do that all right.'

'I wonder, but we'll soon see. Can you open that gate?'

The team had stopped at a barb-wire fence across the trail. I got down, and after a few moments' fumbling, found out the trick of the fastening, opened some four yards of the fence to let the team through, and closed it behind them. Presently Henderson pointed out a wooden shanty some fifty yards away from it. 'George's,' he said, and shortly afterwards we drove between some scattered farm implements up to the door of the shanty.

'Out, I reckon. Hop down and make sure.'

The shanty gave no sign of life as I approached it and knocked on the door. 'Lord! Don't knock. Open the door and go in.' I did so, and the interior seemed to be tenanted by flies only.

'No luck,' said Henderson. 'Here, chuck your hand-bag inside, and we'll go and find 'em.'

'Now where?' he went on, as I rejoined him. 'Can't be cutting, 'cause there's the binder. Buggy's in the shed, and the wagon, so they ain't away. Listen! That's a post hammer. I know, George is fencing a pig pasture. Saw the fence on his wagon two days ago. They'll be t'other side of the bluff.'

He drove round the wood which sheltered the shanty and stable from the north-west, and in a few minutes pointed down a gap in the willow scrub. 'There they are.' Some two hundred yards away I could see two men. 'You're O.K. now, Blanchard. I won't stop as I'm busy. Tell George I'll ring him up to-night about threshing. Good luck.'

I thanked him, and scrambled through the wood towards the two figures. As I drew near they looked to me like a couple of very ragged tramps. 'It's Blanchard, I guess?' said one with a smile. 'Yes,' I said. 'Good! I'm Hartley. This is Gordon.' We shook hands.

'Who brought you out? Pop Henderson? Why didn't he stop?'

A CANADIAN INTERLUDE

'He said he was too busy, but that he would ring you up to-night about threshing.'

'Busy be damned. He knows Gordon's cooking. You'd better take Blanchard back to the shanty, Gordon, and get supper. I'll be along shortly.'

I found Gordon easy to talk to, as we walked back to the shanty. Evidently he had not shaved for some weeks, and sported a black moustache and pointed beard, which gave him a Spanish appearance. He wore a faded khaki shirt, and a pair of bib overalls. These latter garments must have got torn in the fencing operations, for about nine inches of bare thigh was showing on one leg, and a large portion of his shirt-tail protruded from the seat. However, he was perfectly oblivious of these defects in his attire, and chattered to me about theatres and cricket in the Old Country quite gaily. I learnt that he was thirty years old, and had been a medical student in Edinburgh. Presently we entered the shanty, and about ten thousand flies rose up to greet us.

'Blast the flies,' he said. 'They're always bad in the Fall.'

The furnishings of the shanty were simple in the extreme. A double spring-mattress bed was fixed in one corner, stove in another, table in another, and a crude washstand in the last. I watched my companion get supper with great interest. Having lit the fire in the stove, he examined the water bucket. It was half-full, with a dozen dead and dying flies floating on the surface. 'I think not,' he said, 'seeing as how we got company like. Fresh water is indicated. Shan't be a jiffy.' He picked up the bucket and vanished. I sat down on a chair and lit my pipe. I began to think that my life at home had certain features to commend it.

A CANADIAN INTERLUDE

He returned with a full bucket, filled the kettle, placed it on the stove, rolled a cigarette, lit it and then gravely contemplated two frying-pans which were hanging on the wall. There was a trail of grease beneath each one, evidently the drainings of years. 'Now which,' said he, 'did I use for fish the other day?' He took down each, and sniffed carefully. 'Got him,' he said. He placed the chosen pan on the stove, and put some grease into it from a small tin pail. 'Fried spuds,' he explained, and then flumped down on his stomach in the middle of the floor. 'Larder and cellar combined,' he gurgled, as he lifted out three loose floor boards, thus exposing a small pit. 'Ups a daisy.' He reached down, and brought up three dishes each covered with a piece of newspaper. 'Dust works through the floor as we walk about, so we've got to keep everything covered up.' He uncovered the dishes and exposed one of cold boiled potatoes, a joint of cold beef in a baking tin, and a plate of butter. He replaced the boards, rose to his feet, put the beef and butter on the table, and tipped the potatoes into the frying-pan. Having chopped these up roughly with a knife, incidentally dropping in some cigarette ash as flavouring, he turned to lay the table.

The beef and butter were covered with flies. 'Hell!' he said. 'Wonder how long George'll be? Better put 'em in the cellar.' He did this, replaced the boards once more, dusted his hands on the seat of his overalls, and resumed his torture of the potatoes.

The table was covered with what was originally white oilcloth, but which now presented a mottled appearance, cigarette burns and ink spots fighting for pride of place. Subsequently I found out that there were four layers of this on the table, the procedure being to nail on a new bit each year. Gordon next proceeded to lay the table. From a rough shelf he produced the necessary utensils, flipped them on the table with the ease of a conjuror performing a well-practised trick, turned to the stove to chivvy the potatoes once more, and then turned and beamed at me. 'All done by kindness,' he remarked, rolling another cigarette. 'Ah, here's George.'

Hartley came in, hung his cap on a nail, and sat down to the table. He grinned at me. 'Bit of a shock I guess,' he said. 'All set, Gordon?'

Down on his tummy went that cheerful chef once more. He hauled up the beef and butter, and dumped it on the table. 'Come on, Blanchard. Sit up. I guess you're hungry,' said Hartley.

I drew up my chair, and George carved off three liberal helpings of beef, fighting and blaspheming the flies as he did so, while Gordon made the tea.

'Spuds forward,' said Gordon, 'all 'ot like.' He brought the frying-pan to the table, decanted its contents on to our three plates, gave it a scrape with a knife, tapped it on the wood box, hung it on the wall beside its fishy brother, and sat down.

I was young, hungry, and the beef was tender, but I couldn't help looking at the joint. It was one twizzling mass of flies. They soon proved too much for George. 'Get the fly-paper, Gordon. I'll teach the bastards.'

Gordon fetched a tangle-foot fly-paper from the washstand. George waved his knife above the joint to disturb the flies, and then Gordon swiftly covered it with the paper, sticky side uppermost. 'That'll larn 'em,' he said, and calmly sat down to his meal, while myriads of flies rushed to their doom.

I laughed. It was too funny.

'If only our respective mothers could see us now,' grinned George. 'Still, we keep fit on it, don't we, Gordon?'

'It's British grit as does it,' replied that worthy. 'There's always something in this damn bitch of a country. When it ain't flies, it's mosquitoes, and if it isn't mosquitoes it's so damn cold that you feel you could put up with both of 'em if it 'ud only get warmer.'

We finished the meal with bread and syrup, and then George went down to the stable, while Gordon washed up, and I dried the dishes. When George returned we smoked and talked, while Gordon made bread.

He had the dough in a bucket-shaped machine on a chair by the stove. Evidently it had risen satisfactorily, as it had overflowed down the side of the bucket. He placed the machine, a Universal Bread Maker, on the table, and turned the handle till the dough was all in a lump on the mixer. Having first floured the table, he scraped the dough from the mixer, punched it viciously for a while, then packed it into baking tins, and put it in the oven. 'Talk about a ruddy marine,' he said. 'I could give 'em points.'

The telephone rang, two long and one short rings. 'That's us,' said George. He talked to the caller for a few minutes, and hung up. 'Pop wants to see me,' he said. 'He's going to start threshing on Monday. Damn! All the horses are out. Still, p'raps I can catch Duke. Don't keep Blanchard up too late, Gordon. Cheero! He caught up his cap, and was gone.

'That's like all these folk out here,' remarked Gordon. 'Walk a mile to catch a horse to go a hundred yards. Henderson's less than a mile away. He could be over there before he'll catch Duke.' However, Gordon proved to be wrong, for in a few moments we heard George ride past the shanty.

After talking for a while, I began to yawn. 'You turn in,' said Gordon. 'This damn bread'll be another hour, and I guess you're tired. You'd better sleep against the wall, then George won't wake you when he gets back. I'm sleeping in the other corner on those blankets.'

'But I don't want to take your bed,' said I. 'Can't I——?'

'No, you can't. I'm off next week to Winnipeg now you've turned up. That's your bunk now.'

I started to undress, opened my handbag, and took out my pyjamas.

'Lumme,' said Gordon. 'Don't they look lovely. We've each got a pair somewhere, but usually we sleep in our day-shirts. It's, ahem, a dirty habit, I'm afraid, but you see, we do our own washing. Oh, I forgot. Your trunk hasn't turned up yet, has it? You haven't got any working shirts, so you'd better use your 'jamas till you have.'

I put on my pyjamas, got into bed and was soon asleep.

I'd had a full day.

CHAPTER IX

I awoke next morning about six-thirty to the scent and sound of frying bacon. 'Ought I to have been up before?' I asked Gordon, as I scrambled out of bed. 'Where's Hartley?'

'Don't you worry your fat about getting to work. I guess that'll happen fast enough. It's all they think about in this country. George is down the stable. Guess he'll be up in a moment.'

'Say,' Gordon went on, as I started to dress, 'you'd better keep that suit for swell occasions. Give the girls a treat like. Your trunk'll be up in a day or two. Ah! Here's George.'

Hartley came in, and between them they fitted me up with overalls and shirt. After breakfast, Hartley filled his pipe, and turned to me. 'We don't use surnames out

108

here much,' he said. 'Can't help it with Gordon. His name's Cedric. We had to kick at that. So it'll be Jim, I guess, from now on. Get me?'

I nodded.

'Right! Now look here; my folks tell me that you're a farmer. Can you do any work—hard graft, I mean?'

I thought of my hoeing. 'I can do any English farm work,' I said.

'You can pitch hay?'

'Yes.'

'Good enough! Gordon, you take Duke and the buggy, and go to Barloe. You know what to get in the grub line, and I want two balls of barbed wire. Look out for Jim's trunk, and you'd better get him some overalls. Come on, Jim. I want to get the loft filled with hay before we start threshing.'

Apart from a short dinner hour—beef, bread, syrup, and flies—we hauled hay until six o'clock. George pitched the load from the stack, and I pitched it off into the loft window. I found that Canadian slough hay was like lead compared to the English variety, but I don't think I did so badly.

After supper, George asked me if I thought I could drive a stook team on Henderson's threshing gang. Apparently this stook team business was important. I wondered, if I found myself unable to do it, whether I should be sent home as of no use to Canadian farming. Still, according to Henderson, it was only pitching sheaves, so I said that I thought I could manage it.

'I reckon you will all right. 'Twon't be such hard work as to-day. Sheaves are easier than hay, but it means pitching from six in the morning to eight at night, with only an hour out. I'm going to drive one team, and it's money in my pocket if you drive the other. Henderson's paying five dollars a day for a team. You're supposed to get a hundred and fifty dollars the first year, aren't you? Well, I'll give you a dollar a day extra for every day on the gang.'

I agreed, and asked what Gordon was going to do.

'Oh, he's off to the Peg. His folks have sent him a hundred and fifty dollars to take a course in salesman-ship, and he's going down to see about it. Most likely he'll do in the money, and forget about the course. Here he is. Sounds as if he'd got a skinful.'

George was a little unjust to Gordon. He was not full by any means. He had only helped Jack Smithers finish off the remains of the pair of boots which had arrived at Barloe with me the day before, and I found out after-wards that it took more than one bottle to fill him. Still, he was merry.

'Got your trunk, Jim,' he yelled, as we went to the door. 'She weighs a ton. Had to rope the son of a bitch on the back. Stand still, Duke, you prick-eared little bastard. Ah, but darling, I loves you. Come to bye-bye with daddy.' And he disappeared with Duke in the direction of the stable.

George and I got the trunk into the shanty, and also some parcels and the mail, in which there was a letter for me from my father. It was a curious letter. First it gave me a lot of good advice as from a sorrowing parent to an erring son, pointing out that I had no safe back-ground in this new country, but that I should have to stand entirely on my own feet. Then there was a lot of general farm news, which brought it all back. I thought of my home, of my dog, of the peaceful Wiltshire meadows. Should I ever wander through them again with my rod? I looked round the shanty. What a mouldy hole it seemed, and what a silly fool I'd been to come away from home to this. But the inherent love of the father to the son peeped out in the last sentence of the letter: 'And may the Great Architect of the Universe have you in his keeping.' It was many years afterwards before I really appreciated this last sentence.

Still, I got very little time to think about my home people or about myself. Gordon went to Winnipeg a day or two afterwards, and then threshing started. The first farm to be threshed out was, of course, Henderson's, and as this was a mile away we had to get up each morning at four o'clock. Gordon's cooking job descended to me, while George got the teams ready. Thank heaven it was only our breakfasts for which I was responsible. We got dinner and supper with the gang at Henderson's, returning home to the shanty each night about ten o'clock.

In that first week I learnt many things. I found that in a few days it was possible to train one's stomach to go without food for long periods of time. Having toiled from six to nine, I felt ready for a snack, but unless something went wrong with the machinery, we worked until noon without a break. By ten o'clock there would be an aching void where my stomach used to be. I thought of the orderly harvest at home. It was almost stately compared to this feverish thing. Here I pitched off a load of sheaves into the self-feeder. I then drove at a trot out to the pitchers, four husky French Canadians. They tried to bury me with sheaves. I arranged them in

111

some semblance of order, balancing myself with difficulty as the wagon lurched from stook to stook. I
thought of the carters at home; two loaders on each
wagon, and a small boy to lead the horses and call out
'Hold tight' at each move. And as I was scrambling
about, nearly up to my neck in sheaves, I would hear
Jean, the boss of the pitchers: 'Ho, fineesh. Machine,
she wait.' And away they would run to meet the next
wagon, while I would scramble along to my reins, and
drive the load to the thresher. Sometimes a large part of
it would fall off on the way, and the next time I went out
into the field the pitchers would pitch it up again, and
Jean would say: 'You lak make work? Yes! We no like.
No!' And I'd feel awful.

The threshing machine, or separator as they called it,
was only a small one. It could thresh about two thousand bushels of oats, or one thousand bushels of wheat
in a day. Full-size outfits, I was told, could do five
thousand and three thousand bushels respectively. But
this one was big enough for me. It seemed insatiable.
By eleven o'clock I was done. I could not go on. 'Twasn't
work, this: it was slave driving. What would happen if
I didn't go on? They'd laugh—these English farmers!
No! Damn them! I toiled on as in a dream. Dinner time
at last, thank God.

I hitched out my team, took them to the stable,
watered them, fed them, and went into Henderson's
house with the rest of the gang, about sixteen of us in
all. They fed us well, and I learnt to eat large quantities
of food in a few minutes, in order to get in a few
minutes' rest before the machine started again.

About four o'clock, a bucket of tea and some biscuits

were brought out to the machine by one of the grain-hauling teams. As your turn came to pull in with a load, you jumped down, dipped an enamel cup into the bucket, and gulped your tea, while Henderson pitched off your load, which took him exactly four minutes. Then on you went again until eight o'clock, when the insatiable monster ceased its clamouring until next day.

After supper, George and I would get our teams from the stable and go home. There were always a few jobs to be done there; water and wood to be brought in preparation for to-morrow's breakfast, and our teams to unharness and see to for the night. And then, bed, glorious bed.

However, by the end of the first week I found that although I was tired each night, I was beginning to acquire a sort of mastery over the work: that I'd got a knack of balancing on my load no matter what happened; that, although my horses were only half-broken, they knew a lot more about the business than I, and also that they were much more important. Men could look after themselves—if they didn't it was their funeral; but horses must be looked after and fed before men were considered at all. And there was no workmen's compensation if you didn't look after yourself either. I also learnt the blessed peace of utter physical exhaustion, coupled with a beautiful mental blankness. 'Jist do your work, three square meals a day, and tobacco, what more do you want?' said Haines, one of Henderson's hired men. He also said one Saturday night: 'that it was a bloody wise old feller who invented Sunday'; and I thoroughly agreed with him.

As a matter of fact, in our case Sunday was not wholly

a day of rest. The cooking had devolved on me, and this was the only day on which to bake bread. My first lot was awful, but we ate it the following week for breakfast, as there was no alternative. George never grumbled. 'I hate cooking,' he said, 'and after you've eaten your own efforts for a bit, you'll find out how to do better.' I despaired at the time, but after-events proved him to be right, and I acquired quite a reputation as a cook before that winter was over.

One night when we returned home, we found Gordon there. He had returned, as George had prophesied, broke to the wide but still cheerful. Having blown the money his folks had sent him for his training in salesmanship, he was now scheming to earn it so as not to let his people down. Accordingly, after much telephoning, he fixed up to go tanking on a steam outfit some twelve miles away. Henderson's was a gasolene engine.

In a day or so he left on a borrowed horse, and our outfit came to George's farm, but Mrs. Henderson continued to cook for the gang, as the farms lay side by side. Whether this was due to the fact that I had become much too valuable to be taken away from the gang, or whether George feared bloodshed if I did the cooking, I do not know. I hoped it was the former reason, but in my inmost heart I knew it was the latter one.

Just after we'd finished George's crop, it snowed about six inches deep. Then it froze up. 'What'll happen now?' I asked George. 'I thought we had four more farms to thresh out of field.'

'Well, we'll just thresh 'em. Stick the racks on sleighs and carry on.'

We did, and I made the acquaintance of the Manitoba

bobsleighs. It was harvesting extraordinary. The stooks were just mounds of snow to look at. As the sheaves were pitched up, frozen ice and snow whipped across the loader's face. It was about zero weather, and the cold made the engine difficult to start. This was worse than the work, as the cold got right into you as you were waiting about. The pitchers soon chucked it, and went back East. This meant that we pitched our own loads. The output of grain was lessened but the toil was increased. It got dark about six o'clock, but the moon-light on the snow furnished enough light for us to carry on till eight as before. And when there was no moon we managed somehow. Henderson fixed up an electric head-light on the engine which shone on the self-feeder, and when you pulled out from the glare of it you were almost blind for a moment or two. Sometimes he'd light a small pile of straw some few yards away from the engine in order to warm himself. At one farm they hauled the grain straight to town, and the grain teams brought back some whisky one evening. I can remember driving in with a load, and seeing three of the teamsters and Henderson hand-in-hand dancing round the fire like devils in hell, while the forty horse-power engine drove the separator unceasingly and unattended.

The separator made quite a good job of the grain under these extraordinarily difficult conditions, but of course small pieces of ice the same size and weight as an oat or wheat corn, were mixed with it. 'Sell it all quick,' Henderson advised the owners. 'When she thaws out in the spring, God knows what'll happen.' It was good advice. Grain is hauled loose by the railways in Canada in box cars holding up to eighty thousand pounds weight, and hundreds of these on the tracks at Port Arthur had to be emptied into the lake valueless the following April.

Threshing came to an end early in December, and we started to haul out our own grain, white oats. As it had all been threshed before the snow, it made a fair price, thirty-two cents a bushel. It had all been threshed into small box-like granaries, each holding a thousand bushels, which were dotted about over the farm. We used to shovel the grain out of these into a wagon box on sleighs, and haul it to the grain elevator at Barloe, six miles away, making two trips daily. I think I should explain that the tops of all vehicles were interchangeable. They were placed on wheels or sleighs, according to the season.

We were busy grain-hauling when Gordon returned. He had got in forty days tanking at three dollars a day. This accumulated wealth worried him. He paid George seventy-five dollars on the night of his arrival, and got him to write a cheque for that amount to a Correspondence College for a course in salesmanship. He would do the cooking for us in return for his board, he said, which would give him ample time for study. Privately, I don't think he was any better cook than I, but he was good company, and we were glad to have him there. It was about forty degrees below zero then, and it was much nicer to come home to a warm shanty. We had fitted up another stove, a large box affair, which never went out, day or night, but it was too risky to leave the damper open at all when we were away all day, and forty below requires a roaring fire to make it livable.

I imagine that Gordon's course was quite a good one

as these things went, but, at the time, I often thought that had the authors heard one half of Gordon's lurid comments on it as he studied, they would have learnt far more than he did. However, in due course he finished it, passed their examination, and received a beautiful first-class diploma with a large red seal. Moreover, they got him a job selling gramophones, and he departed to Winnipeg shortly after Christmas.

Before he went he took me to Beaver Lake in order to get our hair cut. Apparently this was an annual event; not the hair-cutting—there was a barber in Barloe—but the trip to Beaver Lake. Grain-hauling was over for the time, and the hired men of the settlement had a collective day off, ostensibly for tonsorial needs, but in reality to get drunk. Barloe was five miles away, and Beaver Lake was twelve—but the latter town was wet. I was rather pleased at being asked to join in this revel, I remember, but George grinned and stayed at home.

We went in a wagon box on sleighs, eleven of us. Haines drove a stallion, Bob, on the near side, and a mare, Ada, on the off. These he cursed fluently the whole way. Having arrived at the Lake, and put the team in the livery stable, we lined up at the long bar of the Jubilee Hotel. 'We'll hit her up once all round before grub,' said Haines. 'Right-hand man shouts first.'

This evidently meant eleven drinks. 'I can't drink eleven beers,' I said to my neighbour in the line. 'Keep 'em short then. Try port and brandy.'

It sounds incredible, but I did so, and towards the end of the round I collapsed. They carried me up to a bedroom, and laid me on a bed. There I spent the remainder of the day, in a semiconscious condition, save for periods of vomiting. Another of the party was brought up in the afternoon, also *hors de combat*. As I lay there I began to realize that I had got just about as low as it was possible. I thought of my people at home, especially of my mother. What would she think if she knew? Oh God! I was going to be sick again. I was, very sick. I wanted to die.

However, they got me home all right, and I was ill for three days. George undressed me, and got me to bed on our return, and told Gordon that he ought to have kept an eye on me. That worthy had consumed more than any one, but he was still, in his own opinion, quite sober. 'Never got a—hic—chance,' he gurgled. 'Blighter was out before we got properly started. What did he want to come out here for, if he can't look after his little self? 'Sides, I ain't a bloody wet-nurse, I'm a firsht-clash shalesman, Georgy, and don't you forget it.'

'Oh, go to hell!'

'Allri', I'm goin', but not to hell, Georgy. I'm goin', I'm goin', I'm goin'——'

He subsided on to a pile of horse blankets. George looked at our two sleeping forms for a moment, grunted, put out the lamp, and hopped into bed.

Silence reigned in the shanty.

That trip to Beaver Lake did me a lot of good. It convinced me that getting drunk was an unpleasant business with nothing to recommend it, and I have managed to keep clear of it since that time.

After Gordon had gone away to his job in Winnipeg, I experienced loneliness and solitude. George got engaged to a farmer's daughter who lived some twelve miles west of Barloe, and during that winter he used to go away to her home most week-ends. When he found that I could be trusted to look after the horses and myself alone, these week-ends developed into weeks, and once he was away for a whole fortnight.

My job in his absence was to tend the horses and cut firewood. I used to go out each day after dinner with the team, and hunt the bluffs for white poplar. The timber in them consisted chiefly of the black variety, but this is no good for firing. I'd get in to supper about six o'clock, and spend the evening reading. Mrs. Henderson would ring me up each evening but apart from this I rarely had any conversation with a soul during George's absences from home. It was not till long after that first winter that I realized that the good lady rang up to make sure that I was all right. The temperature varied from zero down to sixty below, and if I had cut myself with an axe, or had been kicked by a horse, or come to physical harm in any way, it might have been serious. I can remember a small cut on my hand festering, and then my arm becoming painful and beginning to swell. I asked Mrs. Henderson one evening as to the best treatment for it, and received orders to get on a horse and go over to her at once. I did so, and she fomented and dressed my arm, and extracted a promise that I would come over each evening until it was well. I needed no coaxing, and was sorry when the cure put a stop to these visits.

We only kept the driving horse, Duke, and one pair of horses in the stable during the winter. The others lived out on the prairie, licked snow for water, and pawed through it to the stubble for food. The water-trough was now a block of ice, which had to be cut with an axe each time you watered your horses. I wore horsehide mittens on my hands for all outside work. These were lined with loose woollen mittens, which were placed in the oven each evening to dry out. It sounds rather horrible, I know, but our socks received the same treatment. Apparently, if you could put them on dry, your feet stood less chance of freezing.

I read voraciously everything I could get hold of during those lonely winter months. Each week I received from home six *Daily Mirrors*, *Punch*, and the local weekly paper. George got the *Tatler* weekly, and the *Windsor* and *Strand* magazines each month. When he was home, we would often invite two other bachelors, who lived in a similar shanty some four miles south, for supper and cards. I don't mean supper proper—we all had that about six when we finished work for the day —but just a snack during the evening about ten o'clock. My mother had sent me a large pot of mincemeat, and I had learnt to make pastry of sorts. Any drawback in its lightness was overweighed by the mincemeat. Henderson came over one night and tried the mince pie, of which he was so enamoured that he went home and told his wife that I was not so niggardly with the mincemeat as some folk he had at home.

In March, the pile of poplar poles, of which I had cut some sixty loads, was sawn up. Henderson brought over an oil engine and circular saw on some sleighs, and we did the job in two days. When this was done, I had a go at boring, monotonous work, as while George went off courting again, I split the heap of logs with an old axe, and afterwards piled them into an orderly pile.

I don't remember feeling at all unhappy that winter,

in spite of the loneliness; but I was glad when spring came along in April. One gets a bit fed up with snow after five months of it. The whole scheme of things was so different to my old life. In that, one told somebody else to do a job, whereas in this there wasn't anybody else to tell. Every little detail had to be done yourself or otherwise it wasn't done. I noticed this most in domestic things. For instance, as a lad I had never realized that salt cellars had to be filled; there was always salt in them at home, but in the shanty it meant that I filled them. Cooking I did not mind, in fact I rather enjoyed it. There was a certain excitement in wondering how things would turn out. But washing clothes was a different pair of shoes entirely. I was a worse washerwoman than I was a hoer. Sunday was our usual washing day, and also the day for bread-baking as a general rule. We had a sort of spring cleaning just before the spring sowing began. It was quite successful for the most part, except when we tried to wash our blankets. George said that they hadn't been washed for two years or so, and he thought it was about time. Accordingly we bubbled them up in a large pot for about a day and a half. Then we took them outside to wring them out. George caught hold one end, I took the other, and we twisted them until all we had left was a blanket rope. They had felted, and were useless, so we had to get some new ones. For good washerwomen I shall always have a great respect. The whole business is a sealed book to me.

But this rough bachelor existence did us no harm. We were always fit, and started on the spring work with eagerness.

CHAPTER X

George's farm was a half-section, or in other words three hundred and twenty acres of land, one mile long by a half-mile wide. As the winter's snow disappeared under the spring sunshine the stubble made its appearance once more. Owing to the late threshing the previous Fall, we had no ploughing done, and were faced with the task of ploughing and planting about one hundred and eighty acres. This acreage was the open land between the bluffs of willow scrub and poplar, which was dry enough to cultivate. As the soil was frozen down to about two feet, the snow water collected in the natural depressions in the landscape, forming small lakes or sloughs.

For a day or two after the snow had disappeared the soil was of the consistency of oatmeal porridge. We spent these days in preparations. The other four working horses, which had been running wild on the prairie all the winter, were caught up, stabled, clipped, and fed lavishly. They were as fat as butter, and this fat melted from them very quickly when they got to work. At the beginning you had a job to get their collars on to them, but in a short while you had to put a thick sweat pad inside the collars to make them fit. Ploughs were overhauled and the shares taken to the Barloe blacksmith to sharpen.

When we did start, I was surprised at the good cultivation, as I had always been under the impression that the Colonial farmer just scratched his corn in anyhow. Generally the ploughed land received three harrowings, and was then drilled and rolled. The soil was like cocoa, and contained scarcely any grit, although large boulders were rather plentiful. A turnfurrow, I was told, would last a man's lifetime. George drove four horses abreast on a two-furrow plough, each furrow being twelve inches, and I had the other three horses on an eighteen-inch single-furrow plough, commonly called a 'sulky'. Both ploughs had seats, for which I was truly thankful before we had been at it many days. There was also a trailing seat on two wheels behind the harrows, and the drill was fitted with a footboard for each wheel, so that we rode on every implement.

We left the stable at 7 a.m., and drove until 7 p.m. We averaged about seven acres of ploughing daily between the two teams, and one man harrowing with four horses hitched to six zigzag harrows was supposed to cover forty acres daily. There is not much more to tell about that year's seeding, save that we began on April the sixteenth, and finished about June the first. However, I feel sure that most people will realize that this plain statement covers a considerable amount of real hard work.

The whole business is a bit hazy to me now. We woke at 4 a.m., we worked until 7 p.m., we supped, we saw to our horses, and we went to bed, repeating the process on the morrow, day after day. On Sundays we baked bread, replenished the loft with hay, washed our clothes, and thanked God for Sunday.

In my innocence I imagined that there would not be a great deal to do after seeding until haymaking, but one does not keep a hired man in Canada for his good looks. George had been corresponding with the owners of a quarter-section of prairie, which lay a half-mile north of our boundary. He finally purchased it for fifteen dollars an acre, and as soon as seeding was finished, we commenced to break it.

All the land in the district had been surveyed by the Government some years before, and divided into one-mile squares, leaving a road allowance of thirty-three yards on each side. The country was thus a chess-board of mile squares called sections. A township was six miles square. The road allowances came out of the acreage of each section. They were not all opened up as roads or trails, only those requisite to the district's needs being so treated; but the others, some of which were ploughed by the owner of the section, were available for the community when desired.

This method of scheduling land seems to me to be an admirable one. Certainly it was very simple when it came to transfer of ownership. There were no complicated deeds. William Jones purchased, say, the northeast quarter of Section twelve in Township seven in Range two, previously owned by Tom Smith, and entered in his name in the Land Registry Office. Transfer the entry to William Jones for a small fee, and the thing was done. I am only giving the broad outline of this

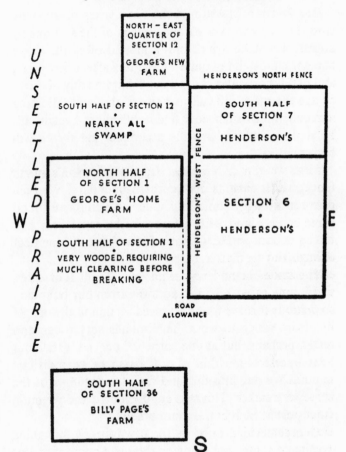

N

UNSETTLED PRAIRIE

NORTH – EAST
QUARTER OF
SECTION 12
•
GEORGE'S NEW
FARM

HENDERSON'S NORTH FENCE

SOUTH HALF OF SECTION 12
•
NEARLY ALL
SWAMP

SOUTH HALF
OF SECTION 7
•
HENDERSON'S

NORTH HALF
OF SECTION 1
•
GEORGE'S HOME
FARM

HENDERSON'S WEST FENCE

SECTION 6
•
HENDERSON'S

W

SOUTH HALF OF SECTION 1
VERY WOODED, REQUIRING
MUCH CLEARING BEFORE
BREAKING

E

ROAD
ALLOWANCE

U N S E T T L E D   P R A I R I E

SOUTH HALF
OF SECTION 36
•
BILLY PAGE'S
FARM

S

business. Possibly there may have been some legal complications and charges, but they were infinitesimal compared with English ones.

Our first job was to find the boundaries of our new land. The surveyors mark the corners of the sections by means of a stake driven into a mound of earth. These become almost, if not quite, obliterated after a few years, and in this particular case we could not find any of them. It was the more difficult as none of the land adjoining was cultivated or fenced. I find it difficult to explain this by writing, but perhaps the accompanying sketch will help matters.

First we ran a line west from Henderson's north fence, which gave us our southern boundary. We then drove in a buggy some four miles west until we found some settled land with a fence in the same line, and traced it back to Henderson's fence—and it came near enough not to matter.

The same method was employed for all four sides, and in one case we had to go away from our boundary six miles before we found a fenced section in the proper line from which to work. This sounds very rough and ready, perhaps, but at one corner of our new estate our lines intersected within a yard or so of an indistinct mound. We dug into this and found the stumps of the surveyor's stakes. This was good enough, and we put a stout post at each of the four corners.

George decided to break the new farm by ploughing north and south, beginning on the east boundary in line with Henderson's west fence, thus ploughing up the road allowance. He got busy marking out the strikeouts with stakes, while I drove four horses abreast on the 'sulky' plough, which was fitted with a special breaking turn-furrow, a very long narrow one.

Virgin prairie is tough, very tough. It lies over behind a breaking plough like endless strips of linoleum, the long turn-furrow laying the furrows quite flat. These strips fitted into each other just as if you had turfed the land with the turfs upside down. It was so tough that in some places, when the furrow broke owing to the plough hitting a stone, I would look back and see fifty yards of my ploughing unwinding back to its original position. When this happened I had to turn the team, go back, and plough that strip once again. I soon found the meaning of the ploughman's saying: 'Run your plough on the near side.' I'm afraid that is a bit too technical for non-ploughmen. It means that your forrow slice should be thicker on the side which flops over in the turn, so that its extra weight will prevent its unwinding on sidling land.

After a week George evidently decided that I could be safely trusted with this job, and he stayed on the home farm doing various odd jobs, and in addition, to my great joy, he took on the cooking. The difference between coming home after a day's work of this character to a supper ready for you, and coming home and getting your own, is immeasurable; it has to be experienced to be appreciated. I used to leave the shanty at 7 a.m., with the four horses hitched abreast to the wagon, in which I would take their hay and oats and my own food. The same hours were worked as in seeding time, and I used to hitch off to return home at 7 p.m., spending most of the middle-day rest of two hours dozing while the horses fed.

A CANADIAN INTERLUDE

That spring George had been given a mongrel collie pup, and most days this newcomer would journey up to the breaking with me. He was an awful fool—I do not remember any other dog in my life so lacking in gumption—but he was company, which was the great thing. He used to follow the plough up and down all day long. Any sensible dog would soon have discovered that I invariably returned to my starting point, and would have lain down to wait there, but not this fathead.

One day as he was following along, a coyote, or prairie wolf, came out of the scrub, and the two played together about thirty yards behind me. I had a ·22 rifle back in the wagon, which I carried across my knees for many turns afterwards until I eventually got a certain chance without hitting the dog, and I shot the wolf.

Later on in the day, I spied some ears pricking up over a mound some sixty yards off, and discovered a den of cubs in a badger hole. Next day I took up a pick and shovel, and dug out four cubs about the size of a small Sealyham dog. At that date the Canadian Government paid two and a half dollars for every prairie wolf killed, and twenty-five dollars for every timber wolf, but these latter were almost, if not quite extinct in our district. It was necessary to produce the heads in Beaver Lake to obtain this bounty, so we put them in a bag, and George took them in a few days afterwards. Years before one only had to show the ears, but apparently some artful individual conceived the idea of digging out cubs, cutting off their ears, and then letting them go to breed more cubs, so in my time heads had to be produced.

I got my bounty for these five heads, but they were never counted. The weather was very hot, and when

A CANADIAN INTERLUDE

George took the bag into the Government office in Beaver Lake, the official besought him to take them out quickly and he would pay anything in reason.

Although a quarter section contains a hundred and sixty acres, I could only find about eighty-five acres of open land to break, of which I got between sixty and seventy acres broken that first summer. It was a curiously shaped field. On only one occasion did I get a strike-out from end to end, as the landscape was broken up by sloughs and bluffs of willow scrub and poplar. Every now and again the plough would stick fast in a hidden root, and I had to haul it out backwards with the team. I always carried a logging chain and an axe on the plough, to deal with any interruptions. Sometimes I would come to a small tree or piece of scrub, hitch off the team from the plough and pull out the obstruction by the roots. The best method of pulling out a tree was to hook the chain as high up the trunk as possible, and hitch the horses to its other end. As they pulled I would see a root move in the ground, and sever it with the axe. Then, I would swing the team to pull in another direction, and repeat the process until the tree came out. Big trees and large patches of scrub were left to be pulled out in future years; the first thing was to bring into cultivation the open land between the bluffs and the sloughs.

I do not remember when I have been more satisfied and pleased with my lot; even to-day the memory of that job gives me great pleasure. In thinking about it now it has come upon me with rather an unpleasant shock, that as far as I can see, I have ploughed my last furrow, as my present farm in England is now all grass,

A CANADIAN INTERLUDE

and I do not possess any ploughs. This thought saddens me, for ploughing is the king of jobs. In itself it is all-sufficing and soul-satisfying. You English townsfolk, who sneer at Hodge plodding at the plough tail, do not realize that he pities you, in that you cannot plough and have never known the joy of ploughing.

'But how monotonous and boring it must be,' you will say, and in the saying you will display your ignorance, for ploughing is the most charming disguise that work can wear. The plough is a perfect implement. The coulter cuts the side of the furrow slice: the share cuts the under side; and the turn-furrow or mouldboard inverts the whole. Therefore, if you are a competent ploughman, you are performing a perfect operation, and since when has perfection been monotonous?

When once you have acquired the knack of it, it goes with the effortless urge of a sailing boat. The plough, which looks so clumsy and uncouth, changes its character. In conjunction with your team of horses, it becomes a glorious galleon, which you steer proudly over the rolling fields like some mariner of old. It is no longer an ugly, awkward, inanimate thing, but a delicately flexible instrument, which responds to your lightest touch.

As you become intimate with it, you find that you have ceased to be the operator of a mere farm implement. You and the plough have become one, a common intelligence with but one idea only, to plough—on and on and on. Your mind stands calmly aloof, rejoicing in a thing in which it has no conscious part, noting with a detached satisfaction the perfect furrow which falls away on your right in an infinite ribbon.

A field-mouse, disturbed by the point of the share,

goes scrambling over the moving furrow, only to be buried alive beneath it. You are sorry for him; you hope he will get out all right; but you are drunk with the urge of the plough and do not stop. Stop? Why, to stop would be absurd. You are no longer a man, you are a ploughman. The mouse must take his chance, and on you sail, unheeding—on and on and on.

Not that the poesy of ploughing is continuous. The length of the lines is determined by the headlands, it is broken into verses by each strike out, and, if you wish to continue the simile, into different poems by the different fields. Such a nuisance these breaks are. Why cannot one plough one long straight furrow for ever without these petty hindrances? But, this being impossible, one is forced to turn, to let the plough grate clumsily along the headland, then to turn again into the work, and swing away on a new tack, happy and interested once more. 'Tis true I am no physician, but I would suggest in all sincerity that three months' steady ploughing would cure any man of a nervous breakdown. For ploughing is a mental tonic of great power. The ploughman is master of the situation. Nothing can stop him. Little by little he changes the surface of the earth. The plough may be slow, but it is so very sure. As the strip of black on the east side of that piece of prairie grew slowly wider and wider until it neared the west boundary, I was forced to marvel at the relentless power of the plough.

Possibly this rhapsody on ploughing will seem absurd to many people, but no one can doubt my competence to sing on so noble a theme. If there be any such doubting Thomas, I would refer him to that quarter-section of land in North-west Manitoba. There have I written my signature with the plough, a signature that will stand when I am long forgotten, a signature of which I shall never be ashamed. And if ploughing generally be conceded a pleasing thing to do, then to plough virgin land is pure joy. The thought that you are ploughing the land for the first time since the world began satisfies your innermost soul. Each furrow is such a definite little stride in the world's history.

That piece of breaking is a thing to which I look back with considerable pleasure, and were it possible I would do it again gladly. But it is not possible, and as ploughing in this country seems doomed, I must be content to have these memories.

CHAPTER XI

I did not quite reach the west boundary of our new farm that summer, as haymaking put a stop to the breaking. That season was a particularly dry one, which meant that we had a much larger crop of hay. This may sound rather Irish, but it is absolutely true. There was very little land sown to cultivated grasses and clovers in our district, and practically the whole of the hay was natural or wild hay.

In the spring the snow water collects in the natural depressions, and the land is ploughed down to the edge of the water at spring level. As the summer advances the water in the sloughs evaporates, thus leaving a lot of dry land, which grows a heavy cut of natural grasses. At haying time you drive your mower round and round the slough until you come to the water line. In a dry season some sloughs dry right out so that you can mow the whole of them, whereas in a wet season the area you can mow is restricted.

Apparently no one in Manitoba knows anything about a hot hayrick. The procedure is to mow one day, rake it up and pook or coil the mown grass the next day, and carry it on the following day, quite green. And my word, it is heavy! We used to stack eight two-horse loads in a day, stacking it in long narrow ricks, made in benches much like an English thatch pile.

And there were only two of us. I think that was the most outstanding feature of Canadian farming as compared to the English type. Only the two of us! When one thinks of the multifarious laborious tasks that must be done, and done at the proper time, on any farm, it seems almost absurd, especially when added to the farm work were the domestic necessities of cooking, washing clothes, mending, and bread-baking. And I had come from a system of farming in which there were men innumerable.

As in most countries, haymaking ran on almost into harvest, and any day not occupied in haymaking was taken up in disking the new breaking: a monotonous, uninteresting job not to be compared to ploughing. It is at this season of the year that the Manitoba farmer dreads a hail storm, for it is possible for a year's work to be demolished in a quarter of an hour. So dangerous was this risk that it was customary for most farmers to insure against it. I have never seen a very bad hail storm, but the following summer a storm swept the lower part of our home farm, and completely hailed out about forty acres of oats, which were cut off as with a scythe, while the farmers immediately south of us who caught the full brunt of the storm lost everything.

We were haying at the time, and hitched out the horses to drive them into a bluff of poplar trees for shelter. When they went into the bluff they were grey, almost white, Percheron horses, but when they came out they were green, owing to the hail driving the green leaves from the trees into their coats. I was wearing a pair of short horsehide gloves, and the top sides of my wrists were raw and bleeding by the time the storm was over. The horses were scared by the thunder and I was forced to hold tightly to the lines, thus exposing a hiatus of skin between my gloves and coat sleeves. It may sound queer for a farm labourer to be wearing gloves for pitching hay, but we rarely did any work with our bare hands. This was due to the fact that the intense cold in the winter made the wearing of lined mittens a necessity, thus making our hands as soft as a woman's. To go straight into spring work with bare hands was asking for trouble, so we wore gloves all the summer.

However, this particular year the hail missed our neighbourhood, although we had several severe thunderstorms. The immediate need to start cutting corn put a stop to haying for the year, and we began the harvest. George drove four horses abreast on an eight-feet-cut binder, which was fitted with a sheaf-carrier to dump the sheaves in fours or fives at regular intervals. I stood the sheaves up in stooks of never more than eight sheaves, so that the stooks would quickly dry out for threshing.

Again George did the cooking and chores or odd jobs, as owing to the heavy dews he was unable to start cutting until about ten o'clock, stopping at 7 p.m. for the same reason. Accordingly, I was able to stook from six until twelve, and from one until eight. Although I 'says it as shouldn't', I was a good stooker, for I kept pace with the eight-foot binder, and most of our oat crop was as long and heavy as any I have seen in this country.

A CANADIAN INTERLUDE

Farther south on the sandy wheat plains the corn crops are light and short in the straw, but in North-west Manitoba they are very heavy, and on new breaking oats will yield over one hundred bushes per acre. I often thought of our 'hilers' at home, ambling gently round the fields conversing amicably as they did so. For me there was no company save the blackbirds, which were innumerable and of all colours, black and red, black and yellow, black and white, red and yellow, and many other shades. But there were one hundred and eighty acres to stook, and no possible hope of anyone else doing it, so it was done.

The Canadian farmer makes a better job of stooking sheaves than any English farm labourer I have ever seen. He drives the butt of each sheaf so firmly into the ground that it will stand by itself, and each stook is built carefully to stand any weather conditions until the thresher comes to the farm, which may be six weeks after cutting. While his own crop is waiting for the thresher the Canadian farmer will have left his farm to work on the threshing gang, and any stooks which blow down have to remain down and spoil.

When cutting was finished, threshing started once again, and I realized that I had been away from home for a year. Threshing went better this year, as we had no early snow, and it was carried out under normal working conditions, but there was another great difference as compared to the previous year. When the gang came to thresh George's crop, I was detailed to do the cooking for them, as Mrs. Henderson was not very well at that time.

Before I describe my efforts in the culinary line, I

would like to pay a tribute to the wives of the farmers in Western Canada. They work. How much they work was brought home to me very forcibly, when Mr. and Mrs. Henderson returned from a holiday in England one winter, during which they had spent a week-end at my home. On their return Mrs. Henderson was telling me of their visit to my people, and asked me what the English farm labourer's wife did all day, especially when she had few or no children.

I replied that I supposed that they did their house-work. 'But what is there to do?' asked the good lady. 'Their bread, meat, and groceries are delivered to their door. Their man goes off before seven in the morning, and does not return as a rule until after five o'clock, while their children are at school all day. Look at my life in comparison. I have a husband and three hired men in the house. I've three children who are all too young to go to school in the winter, and for five winter months it is too cold to let them play outside at all. I grant you I've a Galician girl to help, but all the cooking is done by wood firing, all the washing is done at home, we bake our own bread, and all the water has to be carried in in buckets from the well. In addition, I have to go five miles to Barloe for groceries and letters.'

I could make no adequate answer to her comparison. I do not think that there is one. The sentence about her children not being able to go out in the winter impressed me the most. It was perfectly true. To let them out even on the veranda unattended for a few minutes was risky, as a small child might pull off its mittens and get its fingers frozen in a very short time. I think that I am as fond of small children as most men, but the thought of being shut up in a house with them for five months rather appals me.

The rough hard life of Western Canadian farming has many compensations, such as hunting and shooting, for the men engaged in it, but for the life of me I can see very few, if any, for the women. The generally accepted idea that dress forms one of women's chief interests did not seem to operate here. Henderson was well off, but, even if he had given his wife a Paris model, there was no opportunity for her to display it. Another lady, who had emigrated from England as a young bride, told me that during her first winter in Canada, when she and her husband lived in a log shanty, she often woke up in the morning to find her hair frozen to the pillow.

The male pioneers may have broken Canada's prairies, but it was the women who made, and who are still making, farming possible in Canada, doing dreary monotonous work, chiefly cooking endless meals for cross, weary men.

This was borne in on me very much during that fortnight's cooking for the threshing gang of sixteen men. How I dreaded a wet day, as even though the threshing ceased for a time, it meant one or two more days' cooking.

Our shanty was eighteen feet long by fourteen feet wide, and contained our double bed and other furniture. When a rough table and benches to seat the gang were rigged up there was no room to spare. I gave them porridge and beef stew for breakfast, roast beef and potatoes for dinner, and beef stew for supper, all three meals being topped off with bread and corn syrup as a sweet. Breakfast was timed for 5 a.m., dinner for 12 noon, and supper at about 8.15 p.m., while tea and biscuits had to be taken to the thresher about four o'clock. I did not bake bread, but purchased that commodity with the meat at Barloe, whither I would journey in the buggy once daily.

And it was hot weather. The shanty was a veritable oven, filled with flies innumerable. By the end of a week I felt that at washing up and peeling potatoes I had no equal for speed if not for cleanliness. I do not expect that anything I did was very clean, but the gang were fit and hungry, and as they only saw the finished product, a little dirt and cigarette ash did not matter. Anyway, none of them died, and what is perhaps more remarkable, I am alive to tell the tale.

On the last evening for supper I achieved one of the greatest triumphs of my life. I made a beefsteak and kidney pudding in a two-gallon bucket, and boiled it all day in a large swill pail. When I heard the hum of the thresher cease that evening, I took the pudding off the stove, and turned it out on a large enamel dish. It was a masterpiece. It turned out unbroken—oh, yes, I had well greased the bucket—and stood in glory upon the dish with a beautiful golden bloom upon its crust. The gang ate every bit of it that night, and its fame spread for miles.

Still, I was very glad when our crop was threshed, and I was back on the gang at other farms. Threshing may have been hard work, but there was something to show for it at the end of the day, whereas in this cooking business, one got no 'forrader'; no sooner were the dishes of one meal washed, than it was time to get another one ready.

A CANADIAN INTERLUDE

Threshing finished, we started grain-hauling once again. and it was about this time that I received a letter from my father, suggesting that it was about time to end up this Canadian foolishness and return home. And I found that I did not want to return home. No! Very definitely I wanted to stay in Canada where I was. But why, I asked myself, and tried to find adequate reasons. I argued the whole thing out one night when I was alone in the shanty and endeavoured to compose an answer which would explain to my father.

The advantages of returning home were obvious and overwhelming. Here I was a farm labourer, living a hard life with a complete absence of amusements and leisure, whereas at home I should have every comfort and a master's position. The chief reasons for my not wanting to go home seem now to be puerile and silly, but at that time they were very real, and I wrote them to my father, who, I believe, did not understand them at all.

The gist of the thing was that I had discovered that I could get a living on my own without any help from my family. That folk in the neighbourhood accepted me as a useful citizen, not because I was Mr. Blanchard's son of Partridge Farm, but simply because I was Jimmy Blanchard. This seemed wonderful to me at nineteen. It seems all the more wonderful to me now, as at forty I am obliged to confess that I have grave doubts as to my ability to get a living.

Another reason for wishing to stay was that I wanted to see the prairie, which I had broken, under crop. I do not think that my father appreciated this reason either, but it will be understandable to anyone who has done a similar thing. They, I know, will agree that to go away and not see the first crop would have been the act of a traitor. Besides, I had already agreed with George for another year at three hundred dollars, and as he was relying on that agreement, I could not let him down. This last reason was probably the only one of which my father approved.

I found out in later years that he had expected the return of the repentant prodigal, and was completely mystified by my refusal to accept the role. He journeyed to George's home to interview his father, and in desperation inquired if I had become entangled with a girl.

'Girls?' said Mr. Hartley. 'There aren't many out there, and from George's letters, your boy seems scared of them. Seems to prefer work. George says he's all right, and doing very well.'

'Then what is the attraction out there? They seem to spend most of their time doing labourers' work.'

'I don't know. My boy's just the same, though. Still, they're well and happy, so we ought to be satisfied.'

And with that my father had to be content.

CHAPTER XII

Everyone who gives advice to young authors stresses the need for a strong love interest in most books or stories. I am sorry that I cannot oblige. George's father was correct in his statement to my parents that there were very few girls in our part of Canada, and also that, up to date, I had shown no interest in them.

As a matter of fact there were four unmarried girls in our settlement; that is in a radius of six or seven miles. Two of these were engaged to be married, and as there were fifteen lusty unattached bachelors in the same district, the remaining two girls had no lack of attention. What chance had a green Englishman in his first year?

The usual practice was for bachelors living in shanties to go to supper on Sunday evenings with their married neighbours, and invariably, if these folk had an English girl living with them as companion help, the bachelor helped her to wash up the supper things. Once when a farmer's wife had a fresh girl coming out from England, twelve bachelors arrived for supper on the Sunday evening after her arrival!

I must confess that female society, or rather the lack of it, started worrying me during my second winter. As I was now an experienced hand George spent longer and

142

longer visits at his fiancée's home, and the lonely evenings in the shanty became very irksome. The newness
of everything had worn off. The pictures in the *Tatler*
and the *Daily Mirror* seemed to annoy as much as
interest. I began to wonder whether there was not something more to life than sitting in a hut in the midst of a
waste of snow, and looking at them. My own company
was not enough. I felt that I wanted to talk to somebody.

Then I became vaguely aware that Mary Macpherson
at the Barloe store was possessed of some attractions.
Presently these attractions seemed less vague, in fact
quite definite. I discovered that I derived great pleasure
from a few moments' conversation with her, when I
called at the store for letters and groceries. Working in
the store she had a nodding acquaintance with all the
boys in the district, and, if her father was busy at the
other end of the building, she would flirt gaily with
most of us.

I can remember making excuses to myself to go to
Barloe as often as possible on the off chance of a word
or two with Mary. For many evenings I debated with
myself also as to whether I dare ring her up. The first
time I tried it was disastrous, as her father answered the
telephone, and I had to order some groceries, which we
did not want, to be sent up by any neighbour who might
be in town. I thought this very clever at the time, and
repeated it on future occasions when the necessity arose,
thinking that I had hoodwinked her father completely.
Looking back on it now I think that he realized the real
reason for those orders. But the many chats I had over
the telephone with Mary that lonely winter gave me
great pleasure, although I cannot now remember what

A CANADIAN INTERLUDE

we talked about. But to a lonely youth of twenty, who
spent day after day and night after night in a one-roomed
shanty with no other human being nearer than a mile,
a girl's voice, even though it was a nasal Canadian
drawl, was a comforting and satisfying thing.

One day at the store I wanted some cups and saucers,
and Mary said that I had better come upstairs to choose
them. I followed her up to the loft above the store,
which was filled with bales of goods. We knelt side by
side to pick out the crockery from its straw wrappings.
There seemed to be no need to hurry about this, I remember. Then I wondered if I dared kiss her. I would
have to be quick about it, as old Mac would query too
long an absence. The loft was dim. Her head was close
to mine as we bent over the case. I can remember that
her hair had a nice scent to it. Supposing screamed.
Still— I kissed her.

It wasn't a very successful effort. Certainly not one
of the lingering sort as described in some books and
depicted on the films, but rather an unsatisfactory peck.
Mary giggled and rose to her feet. I followed her downstairs with the crockery, feeling a bit sheepish and also
a bit of a Don Juan.

That was my sole escapade in this direction during
my stay in Canada, for Mary got engaged to a man in
Winnipeg the following summer, and got married shortly
afterwards.

Now Mary was Canadian born, and at that date the
English settlers and the Canadian-born ones did not
mix easily. Whether this is still the case I do not know,
but it was understandable at that time for the native-
born man was then the most narrow-minded individual

144

that it has been my fortune to meet. The English immigrant had, at least, some knowledge of a country other than Canada, whereas the other man, apart from an occasional trip to Winnipeg, had seen nothing other than his immediate neighbourhood, which he imagined was all the world. I should imagine that the War has altered this greatly.

Accordingly Mrs. Henderson spoke of Mary as 'a gum-chewing little horror'. She chewed gum, it is true, as did most Canadian girls, but I will not have it that she was a horror. Gum-chewing in itself is not an obnoxious habit, although to find your partner at a dance chewing against your chest like a ruminant cow is a strange experience. Anyway, Mary made that winter much more pleasant for me, for which I shall be for ever grateful to her, and I write her down in my memory as a good sort.

That winter was easier for me in many ways; I could do my work with one hand, so to speak, and I had made many friends during the year. Other bachelors would drive over in the evenings to smoke and yarn, and sometimes Henderson would honour me with his company. He was very interested in our English farming after his visit to my home, and liked to hear all about it. I remember telling him one night that we rarely threshed out of the field, but stacked everything, and threshed it as we required the money. This sounds unbelievable today, but it was true enough in my boyhood, and it impressed Henderson greatly.

There was also our local dancing class, which functioned in the settlement schoolroom on Tuesday evenings. There I learnt to waltz, and also to perform many and weird dances, one of which, I remember, was called the Old Log Cabin. We had only a piano usually for music, but on the two occasions when we held a proper dance, an old farmer used to play the violin, rather indifferently in the early part of the evening, but better and better as the whisky warmed him.

The square dances necessitated the services of a 'caller off'. This individual gave directions and a running commentary in a sing-song nasal voice. Bits of this I can still remember. 'Lady round gent and gent don't go. Lady round gent and gent al—so. Swing yer corner, lady. Gee old Bill So-and-so's stepping high and wide like a colt. Hey! Allum and left.'

I am afraid that I was a poor performer, and I was never sure what 'allum and left' meant, but I think it must have been 'hand in the left' for the grand chain.

I was always a mystery to me why we never got pneumonia at those dances. The schoolroom was always hot and crowded, and we danced in fairly heavy winter clothes. Smoking was taboo inside the room, and we used to slip outside for a cigarette without putting on even a hat. Perhaps it would be forty degrees below zero outside, and in a few moments one's hair would be frozen quite stiff and crackly, but this never seemed to have any ill effect.

Towards the end of the evening the bachelors of the company, who were to drive some particular damsel home, would fill the oven of the school stove with large stones. At the close of the festivities these were wrapped in blankets, and placed at the feet of their passengers. I suppose that this crude form of hot-water bottle made it just possible for young love to go the long way home in spite of the cold. Large family parties drove to these dances in a grain box on sleighs, but couples came in cutters.

The summer roads of Western Canada were just dirt tracks, and we got a much better road in many ways during the winter. For one thing the snow trail was smooth, and the sleighs ran along almost silently. Another feature was that the winter road was more direct. For instance, when we went to Beaver Lake in the summer the road wound round the north end of the lake, but in winter you could drive straight across the lake, some two miles, even with a load of grain, the snow-covered ice making a perfectly level trail.

The horses drawing sleighs walked in the track of the previous sleigh, and when only one horse was driven in a cutter—a vehicle like a large arm-chair mounted on steel runners—the shafts were attached to the near or left-hand side so that the horse trotted in the runner track.

Actually the horses could not possibly get along anywhere else than in the runner tracks. The intense cold made the snow of the consistency of castor sugar, and the slightest breeze caused it to blow about; it would not even stick together to form a snowball. Consequently, the track made by a sleigh and horses got blown in with loose snow soon after it was made. The next tracks suffered a like fate, and so on, the result being that in a few weeks the road was composed of two strips of hard packed snow, some two feet wide and two feet apart. These got higher and higher as time went on, and by Christmas, if a horse stepped off the trail, either outside or in between these two strips of packed snow, he would sink down sometimes to his belly in soft powdery snow.

Of course, the horses got to know this, and needed no driving on a good trail. This was a blessing, not only to the squires of dames, but also to the teamster, when he was hauling grain in a very cold snap. He would be able to hitch up the reins, and walk behind his load to keep warm. This was why the green Englishman on a large farm was always given the young horses, as they had to be driven for some time before they acquired a trail sense.

This may have been a harsh method of education, but it was usually quickly efficacious. The youngster sits up on his load to drive his team, and rapidly freezes. His hands, encased in wool-lined fingerless mittens, soon become devoid of all feeling as he grips the reins. His feet cease to belong to him. His nose is not, and his forehead becomes a dull ache. At the end of his journey his mittens are frozen into hard shells, and he is unable to bend his hands to unhook his team. But after a few days of this he is able to leave his team to their own devices and plod behind his load in comparative comfort.

In the early part of the winter all the grain was hauled 'loose', being shovelled into the grain box from the granary, and from the grain box into the elevator pit at the railway. Towards the end of the winter the grain was bagged into two-bushel bags, as when milder days came, the trails would not carry the weight and loads got upset. By this time the trails were very high in places where they had been continuously blown in; some of them would be three feet high, and when one side collapsed under the weight of the load, you had to heave your load off before you could get your sleigh back on to the trail. For this type of hauling we used a flat platform on the sleighs to facilitate loading and unloading. I have had to unload and reload three or four times in a five-mile journey, when hauling grain in late March. Presumably this is one reason for the grain being put into small sacks of only two bushels. Imagine one man struggling in the snow with our absurd sacks of four bushels!

I find it difficult to convey by writing the fact that the snow ruled our lives for five months in every winter without a break. From October to April it covered everything. We never saw the soil, the roads, or even the sidewalks or pavements in the towns. On a soft day in March you would see a small child being pushed along the side walk in a perambulator on runners.

Our shanty had a one-slope roof, and the door was in the middle of the high side. If you opened a tin of salmon, say, and wanted to dispose of the empty tin, you opened the door, and lobbed the tin backwards on to the roof, down which it would roll to the ground on the other side. When the snow melted in the spring, all your sins of this character, which had worn a snow-white mantle for so long, came to light, and there would be a hideous heap of the most horrible rubbish behind the shanty: empty tins, old socks, meat bones, prairie chicken carcasses, and all the discards of your winter's life. This had to be cleared away before the summer sun got busy.

We got fat during the winter, as we did less work, and owing to the intense cold ate the more. I was over

six feet in height, but in the summer I rarely weighed
over eleven stone, while in the winter I always went up
to over fourteen. The intense cold was, I think, the chief
reason for nearly all the men being clean-shaven. If you
had a moustache your breath froze on it, forming icicles,
and at times ice also formed on your eyelashes. I have
seen men so adorned come in from work, sit before the
stove, and pull large pieces of ice from their moustaches
before they sat down to dinner. It was not a pretty
sight.

We smoked plug tobacco, and I have never found
any tobacco to equal it since I left. To enjoy a pipe to
the full, you must go to Western Canada..Occasionally,
we would get a week of 'sixty-below' weather, which
meant an entire absence of wind. After a good breakfast, with half a gallon of oatmeal porridge, say, as a
starter, fill your pipe carefully, light it, pick up your
axe, and step out into that white, cold, still world. That
pipe is superb. Your puffs of smoke will ascend verti-
cally to the heavens, the flavour is ambrosial, and the
crunch, crunch of your moccasins on the snow is a
pleasing accompaniment. Smoke carefully as you walk
to your work, but keep it going, or otherwise your pipe
stem will freeze up. When it is finished do not tap it out,
but keep it in your mouth all the morning as you work.
A pipe stem is a comforting helpful thing to bite on each
time your axe bites deeply into the tree. And when you
return to dinner you will find a rim of ice around your
pipe stem, so lay it on the back of the stove to thaw out
ready for you after your meal.

Snow in England means to most folk tobogganing,
snowballing, and other pleasures, but I never saw winter

sports of any kind in Canada. I suppose it was too cold. In short and rare doses snow may be an amusing thing, but when you live surrounded by it for five solid months you are something like the man who was sick of the palsy, for verily you are sick of it.

CHAPTER XIII

This record of my Canadian farming seems to be almost entirely about work. Of course, Canada is a country in which the white man does work, and works hard; there is no native population to do it for him as there is in some other colonies. In fact, in my time, the dwindling natives did no work, or precious little. They lived on Government reserves, and were offered every inducement to farm, by such bribes as free seed corn and implements, but they did not exert themselves very much. During some periods of the year they were allowed off the reserve, and travelled about trapping musk rats and other fur-bearing animals.

But most of the white men in the district contrived to fit in a good deal of sport during the year. Wild duck were plentiful until freeze-up, when they went south to warmer climes. The prairie chicken, a species of grouse, and the bluff partridge stayed with us the whole year. At that date one could shoot almost anywhere. The best method was to drive in a buggy, and shoot over anyone's farm, where you found chicken or duck. The partridge rarely flew, but would remain perched on a bush until you almost poked him off, and then he went like greased lightning.

In the winter the prairie chicken congregated round the straw piles of the threshing sites. We had a pile exactly ninety yards from the shanty door, and had got the range off to a nicety with a ·22 rifle. There were occasional jack rabbits, an animal like a large hare, and the bluffs were full of a smaller variety of rabbit. These last changed colour in the winter for protection from coyotes, being brown in summer, and changing to white when the snow came. One season, I remember, they made a mistake, and adopted their winter fashions too soon. We had an early fall of snow, which only lay for about ten days, during which time the bluff rabbits went white. Then we had a short Indian summer for a week or so, the snow melted, and every rabbit showed up distinctly against the brown background.

They were not very good eating, but we went out one afternoon that time, and shot about fifty, which we skinned and placed in an open box on the shanty roof out of the way of the coyotes, a perfect cold storage. We did the same with a lot of mallard duck another year. There was a large marsh about six miles to the south-west, and just before freeze-up the only patch of open water in the whole district was in the middle of this We must have timed it to a nicety, for on the night that six of us went down there, all the duck in the neigh-bourhood were coming to this water, and the next day they all went south. We sat around this water with our boots wrapped in straw bands for warmth, and shot hundreds of duck in the moonlight. I can remember our share of the bag was exactly fifty-seven, as we plucked them the next day, and put them on the roof. This was in October, and we ate fresh wild duck in March.

One part of a bank manager's job was to go shooting in the fall. By this means, as he drove from farm to farm, he got a pretty accurate idea of whether the particular farmer was working his land or not. This was important, as the whole country was run on credit. A young bachelor, who had worked steadily in the district for a year or two, might purchase a chunk of prairie for from ten to fifteen dollars an acre, paying one dollar per acre down, and leaving the remainder of the purchase price over a period of ten years at six per cent, paying off an annual instalment of the loan. His horses and implements might be purchased on the same basis over three years. It was very necessary for the bank, which financed this sort of thing, with the character of the young man as the only security, to know whether he was getting down to it, and breaking the prairie.

English banks have to me an almost church-like atmosphere. One enters them reverently. One is ushered into the manager's office as into the holy of holies, and personally, I always associate an English bank manager with striped trousers, spats, and a grave seriousness of manner like a doctor.

His Canadian counterpart was very different, in speech, in dress, and in habit of mind. The Beaver Lake manager drove up one day about eleven o'clock, and found us busy getting a load of hay from a hayrick. 'Say,' he drawled, 'what time do you boys have grub?' We told him at twelve o'clock. 'Guess I'll shoot around a spell, and feed with you.' He arrived at the shanty before we did, and when we went in to dinner, he had the stove going, the kettle on, and a pan of potatoes on the fire.

At that date horses were the only means of transport, and I learnt to ride as a matter of course, and to ride well. I do not mean that I had a good seat on a horse or anything like that, but I could stick on anything, and make it go where I wanted, which was the main thing. In seeding time one year, the only horse not hard at work was an unbroken two-year-old, and I rode him round the settlement that summer with no saddle, and only a rope round his neck. Anything was preferable to walking.

One of our neighbours to the south of us, Billy Page, had three wolfhounds, and we used to go wolf-hunting in the winter. We had to choose the right day, or the hounds had no chance. When the snow was hard enough to carry both hound and wolf, or when it was soft enough to let them both through as they ran, were the only possible opportunities. Most of the time the hounds would go through, while a wolf could run lightly over the top, and laugh at their efforts to catch him.

I do not think that we were very popular when engaged in this sport, as we carried wire-cutters, but we always went back and repaired any damage the following day. Billy was a harum-scarum individual, who would take any risk. Once we ran a wolf to ground into a short badger hole. We dismounted, peered down the hole, and could see the wolf's head—he had turned round —about eighteen inches down. To dig was impossible, as the soil was frozen hard for two feet down, and as we were miles from any house we could not get any tools.

'No go, Billy,' I said. 'Guess the blighter's done us.'

'See him in hell first. If that God damn son of a bitch thinks he can get away with that, he's got another guess coming. You mind the horses and watch he stays put, and I'll fix him.'

I stayed by the hole with the three hounds scratching and yowling at its mouth, while Billy disappeared into a bluff near by. Presently he came back with a piece of straight willow, about two feet long and as thick as a man's wrist. He lay down over the hole, and worked the stick into the wolf's jaws as a gag. Then he reached into the hole with both hands, keeping the stick tight up against his chest, gripped the two ears of the wolf, drew him out, and tossed him to the waiting hounds. I would not have risked it for all the gold in the Klondyke.

On occasions Billy would go to Beaver Lake, and return home rather merry. He always got home all right, as Blacky, his horse, would stick to the trail unattended. His road home ran through our farm, and one summer we fenced across it, and opened an alternative route in a different place. We were fencing a new horse pasture, and one of its corners came near the old trail.

Billy had gone to Beaver Lake that morning by a different route, and had attempted to return through our place. I was alone that night, and about twelve o'clock, I was awakened by Billy blundering into the shanty.

'What's up?' I asked.

'Doh know,' he mourned. 'Comin' home from the Lake, shtruck wire. Turned west, shtruck wire. Turned south, shtruck wire. Bloody well fenced in.'

I laughed and lit the lamp. 'We've altered the trail,' I said. 'Wait a bit while I get some clothes on.' But Billy was asleep on the floor, so I covered him up with a horse blanket, and went out to find Blacky tied to our new fence. I drove the buggy home, put Blacky in our stable, and returned to the shanty to find Billy still snoring. He was too heavy for me to get him into bed, so I left him there until next morning.

Deer-shooting was a very short season, only about a fortnight in December. Four of us, the best number for a party, went in a sleigh one season up to the bush in the Riding Mountains some twenty miles north of Beaver Lake. This was the only occasion when I made good use of my white tennis trousers and sweater. You had to wear white in the bush for two reasons. One was that it gives you a better chance to get near your quarry, as you always have a white background of snow, and the other was that if you wore any dark colour some silly ass would very likely shoot at you in mistake for a moose. It was also best to take a pair of white horses for the same reason. One man took up a grey and a bay one season, and as he was watering them at a muskeg hole one morning, someone shot the bay for him. When we got up to the bush we bagged a log hut made by some previous hunters. We were lucky to do this, as otherwise we should have had to build one ourselves.

This log hut was large enough to accommodate all of us and our team of horses. As a general rule we took it in turns to stay in camp and cook, whilst the other three hunted in the surrounding bush, using a pocket compass as a help to find their way back to camp. Usually we hunted in singles, but as I was a new hand at the game I went with Billy Page, or I should probably have lost myself. Once we returned to camp empty-handed to find that Haines, who had been cooking in camp that day, had shot a moose, which had walked through the camp clearing. I was told that this was not an uncommon thing to happen, as the hunting of the other three in the country around the camp disturbs the deer, who may then walk near the camp.

Deer shooting was fifty per cent luck, and fifty per cent refraining from smoking. There is rarely any shooting skill required. The two deer I shot were both killed at about fifty yards range. We scouted around until we found a fresh track, which we followed in hope. As our white togs rendered us almost invisible, we were able sometimes to spot the deer before they spotted us, but if we had smoked we should never have got near to anything. It was necessary to take our rifles to pieces and wipe all the oil off the locks and striking mechanism. If you omit doing this when you pull the trigger nothing happens. The oil will have frozen up, and the striker cannot fall.

There were dense swamps of tamarack trees in the bush, the trunks being so close together that a moose had a difficulty in getting through them owing to the width of his antlers. As it was absolutely still, if you stood quietly, you could sometimes hear the 'tap tap' of the deer's antlers against the tree trunks as he wound his way through them perhaps half a mile away.

During the day's hunting you might see one of your companions working his way through the bush following a track. If you placed yourself in his line of advance, and stood perfectly still, he would in all probability blunder right into you before he saw you, more especially if you held your rifle behind you. Apart from your face, your white clothing made you a part of the landscape.

A CANADIAN INTERLUDE

There was only the one room in a bush hut, which did duty as kitchen, dining-room, bedroom, and stable for the horses. In the evenings we played bridge with the horses contentedly munching hay near by. In one seat at the table you had your back to the horses' heads, and they would often cease their munching to gaze tranquilly at our gambling. When I cut that seat I always thought that they were mildly astonished at some of my calls, as doubtless was my long-suffering partner. In after years I have many times wished a human spectator of my card play into a warmish region, but I never minded the horses' sweet-smelling hay-scented breath behind my shoulder.

There were both moose and elk in the bush and we got specimens of each. The law allowed only bulls to be killed, and the Game Warden examined all sleighs as they returned from the bush. I never considered that it was a good law, and certainly it did not have the desired effect, as it led to cows being killed just the same. Bulls were tough eating, but of course carried the desired head, so you killed the first cow you saw, and trusted to luck to get a bull later on, so as to have his head to put with the cow meat.

I cannot remember any other definite form of pleasure, but we always seemed to be having fun of some kind or another. I suppose that when one is young and fit, one does not require many outside aids to merry-making. The joy of life is in you, and finds expression no matter where you may be. Very certainly, I was never so care-free before I went to Canada—boarding-school is a serious business—and I can truthfully say that I have never been so free from care since that time. Everything

159

was funny. You were funny. So were your neighbours, your horses, and your other animals. In summer the mosquitoes were funny, and in winter the cold was funny. Anyway, that was the best way to look at it.

We kept a small pig in the stable one winter, having railed off a two-horse stall as a sty for him. At one point in the outside wall of his quarters there was a crack in the boards. This was the piglet's first and only experience of a Canadian winter, and he was rash enough to sleep one night with his tail against the boards near this crack. Next morning his tail was frozen to the side of the stable, and we found him running round the pen in great indignation with a red dot where his tail should have been. This never healed over, no matter what ointment we used, but glowed like a rear light until he came to his appointed end.

Of course, part of this carefree feeling of which I have written was due to the fact that the average bachelor was responsible to no one. The country was sparsely populated, and as a general rule one's actions only affected oneself. You did not worry about what other people might think. If you wanted to do a thing you did it, and humped the consequences. And I do not remember that any of us thought about consequences at all. Whatever happened you could only let yourself down, family considerations being non-existent.

One fall after threshing was over, Billy Page and I drove about fourteen miles west to a farm where the farmer was selling his turkeys and pigs, by means of sweepstakes on pigeon-shooting and target competitions. There was a large crowd and we did pretty well. Billy won a pig, and we got nine turkeys between us.

A CANADIAN INTERLUDE

There was very little room in the buggy for us by the time we had got all our spoils aboard, all alive-o, but we squeezed in, and set off for home about 11 p.m.

About half-way home the trail ran close by a stable belonging to a friend of ours named Ernest Hudson. It must have been nearly one o'clock and I was nearly asleep, when Billy pulled up.

'Say, can you get hold of the rifle? Hudson's turkey's roosting on the stable roof.'

'Don't be a damn fool, we can't haul any more. 'Sides, Ernie's quite likely to loose off at us if he wakes up.'

'Gimme the rifle. Ernie'll never know who did it till we tell him.'

I fished out the rifle, and took the reins. I had a job to get Blacky to stand still, as it was coldish and he wanted to be home, but presently the rifle went off, and so did he at full gallop.

When I got him steadied down, I asked Billy if he had hit the turkey. 'Bet your sweet life. Plumb centre. Hark at Ernie's dog. I guess we'll slide.'

We slid.

Two or three days afterwards we received a bill from Hudson through the post. It ran like this:

Messrs. Page and Blanchard.
  Dr. to
    Ernest Hudson.

To shooting one turkey	3 dollars
To shooting same without permission	30 —
To waking me up	300 —
To penalty if you do not come over on Friday and help eat the said turkey	3000 —
	3333 dollars
This can be spread over 33 years at 33 per cent	

A CANADIAN INTERLUDE

We went over on the Friday, and took him two live turkeys and a bottle of White Horse whisky.

All this sort of thing may seem childish, but at the time it was jolly good fun.

CHAPTER XIV

Although I do not want to make this book in any way a technical one, I think that some description of the system of farming in North-west Manitoba should be given. Unlike most types of farming, it is possible in this instance to begin at the actual beginning, that is, at the virgin prairie.

The first year this was ploughed, and disked in the summer, and left in that state all the winter. The following spring it was harrowed to a fine tilth, and then drilled to corn, usually oats. Towards the end of my stay wheat was becoming more popular for new breaking in our district, as the new variety, 'Marquis', was coming into fashion. This matured a week to ten days earlier than 'Red Fife', which was the general Canadian variety.

This earlier maturing made all the difference, as we were far enough north to make the growing of Red Fife a chancy business. If frost occurs before wheat is cut, it yields only a poor sample of chicken feed, which is valueless for milling. I should imagine that the Marquis

163

variety widened the wheat belt twenty miles farther north for a considerable distance across Canada.

Generally speaking, the rotation was Prairie, Breaking, Wheat, Oats, Oats, Oats, Barley, then Summer Fallow, and repeat the six corn crops again. There was some land in our district which had been cropped under this rotation for thirty years, and still grew good crops. The summer fallow was necessary every seventh year, not only as a rest for the land during which time it could regain fertility, but in order to destroy weeds.

The wild oat was the chief bane of the farmer. As in this country, it was shed out of the head on to the ground before the tame oats were ripe enough to cut, and in the fall it would not germinate, but lay hidden quite safely under the snow all the winter and came up with the corn crop the following spring. The only way to reduce them was to summer fallow. In July, the last year's stubble set aside for fallowing was waist high with wheat, tame oats, wild oats, and pigweed. This made a good green manuring when ploughed in.

Although grain formed the bulk of the farming, poultry, pigs, and dairying were increasing amongst the married farmers. The lack of these was the one great advantage in working for a bachelor, as all the work in connection with these other branches of farming was extra to the grain farmer's day. On a married man's farm you worked the usual hours in the field from 7 a.m. to 7 p.m., doing this other work in the very early morning and after supper. Besides, these things had to be done on Sundays as well. I have seen Henderson's hired men, after a day's ploughing until 7 p.m., go out after supper to get the cows in from the pasture. After milking, the milk was carried up to the house and separated. The skim milk was then carried back to the yard and then the pigs and calves had to be fed. Presumably they had the remainder of the day for recreation. The cream was sold to a creamery in Beaver Lake, being sent in about three times weekly. Of course, one got better grub with a married farmer, but against that one was rather more definitely the 'hired man'. It was impossible to preserve that relationship between two bachelors living in a one-roomed shanty.

Pigs and poultry went together always. Hens laid eggs all right in the spring and summer without much attention to feeding, but the great difficulty was to keep them warm in the winter. Accordingly, the fowl house was usually a loft immediately above the pigsties. In the sties you put as many pigs as would go in and then a few extra, pursuing the same stocking policy with the poultry. This may sound unhealthy for both, but it was very necessary. Of course, I am writing of pre-war conditions, and possibly some better means has been devised by now. Anyway, if a hen stayed outside the house at night in the winter, and roosted on a bush, her feet would freeze before the morning.

Work was king, men being of secondary consideration. Machines and horses cost money, and must be kept in repair or rested in due season, as otherwise they would break down. But men's powers were presumably elastic, and their endurance could be strained indefinitely, the only limit being when the job was done. And the men never broke down or gave up. A man who did so would have been considered a 'poor tool', and we were all too proud to acquire that reputation.

A CANADIAN INTERLUDE

Nearly all the farmers were owner occupiers. It is true that many of them were not the absolute owners, but were paying for the land in yearly instalments, with the fairly certain prospect of owning their farms in due course with average luck. There were few tenant farmers, and these rarely paid a cash rent, but rented 'on shares' with their landlord. In such cases the landlord took one-third of the grain crop as rent, the amount being determined on the thresherman's tally of bushels. Any other return from dairying, pig-keeping and the like was the sole property of the tenant.

Threshing from the field cost per bushel: five cents for oats, six cents for barley, and seven cents for wheat. For this charge the thresher owner provided the pitchers, the wagon teams, and threshed the grain into the farmer's wagons or granaries. All the grain passed through a Government stamped weigher, which mechanically recorded the number of bushels before they left the machine.

As far as I can remember, oats at under thirty cents a bushel, and wheat at under eighty cents, meant poor times for the producer, while forty cent oats and dollar wheat meant prosperity.

Looking back on that time I have come to the conclusion that the chief compensation for this continuous toil was the fact that the country as a whole considered that farmers were important people. Without farming Canada would have been nothing. If you went into a shop in Winnipeg, no matter what kind of shop, the proprietor would ask you quite seriously as to the crop prospects in the West. All the townsmen knew that they depended, and that they depended entirely, on the success and prosperity of the Canadian farmer. Consequently he considered, and quite rightly, that he was an essential factor in the scheme of things, which gave him self-respect. There is an interesting contrast in the attitude of the British public to-day to the British farmer. Quite a large proportion of them look upon him as an unnecessary nuisance.

I did not know anything about the politics of the country and was not interested, and on the one occasion I voted at an election my vote was purchased at the price of a half-day's holiday. I was ploughing one morning on our home farm during my last year in Canada, when a motor car of all things came along the trail near by and stopped. The driver, a burly man of about sixty, came across to me, and explained that as I was twenty-one and had been two years in the country I had a vote. He told me that he was the Liberal candidate, and inquired my politics in England. I said that I was a Liberal, and he said that I should of course be twice a Liberal in Canada. I pointed out that I knew nothing about Canadian politics, cared even less, and also that I was working for George Hartley and therefore could not get away to go to Barloe to vote. 'That'll be all right,' he said. 'I've seen George in the Lake this morning, and he said you could knock off to vote. I'll drive you in, and he'll come round that way this evening and bring you back.' This was an entirely different proposition. I would have voted for the devil himself, if by so doing I could take a half-day off. 'But I'm filthy,' I said. 'Never mind, I can wait while you clean up. Say, let me take 'em, and finish the turn.'

I was driving five horses abreast on a three-furrow

walking plough, and the old boy took on like a master. It was probably good electioneering for him to do so, but he knew all about it, and finished the round in fine style. While I was putting the team away, he went up to the shanty, lit the fire, and had some shaving water ready for me when I arrived. I think that almost everyone in Western Canada had been a farmer at some time or another. This man had, and as he drove me to Barloe, he told me that he would prefer to plough all day rather than go electioneering. I voted for him, and he got in by a large majority.

One winter, Billy Page, Joe Haines—who was working for Henderson—and I went to Winnipeg for a few days' holiday. Our combined ages totalled only sixty-six years, and we were hungry for town, bright lights, and crowds, after two years' work with a complete absence of these delights.

That holiday went the way of all such holidays. We feasted royally and expensively. We bought clothes, especially ties, which we wore on our return about three times a year. We took in the theatres and cinemas. We had one glorious evening at a boxing exhibition in the Walker theatre. We bought absurdly expensive presents for Mrs. Henderson and her kiddies, and weird and wonderful pipes for Henderson and George. We luxuriated in shaves, haircuts, baths, and face massage. We smoked cigars which we disliked, used hair oil which we loathed, and for some unknown reason purchased some real estate lots in Alberta. I shall always remember that real estate office because of a notice displayed in it. It went like this. 'Please do not swear in this office, not that we give a dam' if you do, but it sounds like hell before strangers.' In short, we had a wonderful time, and spent most of our money.

On our last evening Billy unfolded a scheme by which we could make our expenses. He had been told about it by an old man in Beaver Lake named Jim Bolton. Bolton was one of the old-time cowpunchers of whom there were then very few. I had seen him shoot with a heavy Colt, and he could keep a tin can jumping with it like a live thing.

I did not much care about this scheme of Billy's. There is a streak of caution, possibly a yellow streak, in me somewhere. I could never do risky things for the sheer joy of doing them as did Billy, although I could screw myself up to it if absolutely necessary. Still, my companions were set on it, so I meekly went with them, with a nasty, sickish sort of feeling in my tummy.

Billy gave us definite instructions as to our respective parts; we took a taxi to a gambling hall, and left instructions with the driver to wait round the corner ready for a quick getaway. I suppose it was a typical gambling joint, but this was my only experience. Faro, fan tan, roulette, and poker, were in full blast in a large hall with about thirty or forty men present.

We ordered drinks, and stood watching the roulette. Presently Billy drifted away, and fetched up at a table where a solitary man was idly playing with three walnut shells and a pea.

'Say, Boss,' said Billy, 'what's the idea?'

'Quickness of the hand deceives the eye, stranger,' was the reply. 'Watch the li'l pea.' Flip, flip, flip. 'Where is she? Betcha a dollar you don't know.'

'I'll go you,' said Billy, pulling out a dollar bill.

169

'Lemme think now. When you're full, choose the middle one.'

'You're quick,' said the dealer, turning up the middle shell and showing the pea hidden beneath it. He handed Billy a dollar and spread the shells out again.

By this time Joe and I had drifted across to the table, and stood one on each side of Billy, who was sitting in a chair facing the dealer. After Billy had won five out of the next seven throws at a dollar a throw, the dealer dared him to have a decent bet.

'I'm the owner of this layout,' he said, 'my regular dealer for this game's away to-night. That's why I'm wasting my time here, but it bores me playing for chicken feed.'

'Well, I got a li'l roll I don't mind adding to,' said Billy, 'but you must put your money up on the table alongside mine, if we're going into bigger figures.'

'Goddlemighty!' said the dealer. 'Whatcha take me for? I got more in my jeans than a farmer like you'se ever seen.'

'Here you are,' he went on, exhibiting a fat roll of bills. 'What'll a piker like you bet? One, two, three, five hundred? Anything you like up to a thousand.'

'Alri',' said Billy. 'I'm a farmer alright alright. Don't get het up. I guess I'm a bit canned or something, but your face don't look near as nice to me as it did a while back. Still, you said it. I'm a farmer, and I'll go you five hundred bucks on one throw, an' remember, farmers ain't so easy as they used to be.'

The two rolls of bills were counted and checked, and the thousand dollars placed in the middle of the table in one roll. The dealer spread the shells out

170

swiftly, and said viciously: 'Now then, farmer, where is she?'

'Serious this,' said Billy. 'Lemme think. Never desert your first love's my motto. She's under the middle shell.'

As he spoke, he reached out and covered the middle shell with his right hand, and grabbed the roll of bills with his left. Simultaneously, Joe's forefinger descended on the left-hand shell, and mine on the right. Before the dealer had recovered from his astonishment, we turned up the two outside shells, to find no pea under either.

Immediately Billy turned up the middle one with the same result, while Joe grabbed the dealer's hand, and forced him to exhibit the pea, snugly hidden between his fingers.

'Hi,' he yelled, 'watch the door. They're bouncing me.'

'No, darling,' said Billy, 'you're just getting the re-bound,' and with a heave he upset the table on top of the dealer and dived for the door, closely followed by Joe and me.

Two men got there before us, and must have regretted the fact for days after. Billy hit one beautifully on the jaw, and as he fell he was trampled in our wild rush through the door into the passage.

Joe smothered the second man in a bear-like hug and bumped him into unconsciousness against the door jamb. Down the passage we rushed, luckily opening the street door without any difficulty, and ran swiftly round the corner to our waiting taxi.

'Let her ramble,' yelled Billy, as soon as we were all aboard. 'We're farmers all right, and we've finished harvest. You betcha!'

CHAPTER XIV

I do not think that this record of my time in Canada does justice to George Hartley, my employer. Most of it seems to be about my doings with other folk as companions. Of course, when one lives so intimately with a man as I did with George, one is rather apt to take him for granted. But he was, without exception, the most kindly, courteous, good-natured gentleman I have ever met. He seemed to me a sort of Rock of Gibraltar on which I could depend in any difficulty. He never lost his temper, he was always fair, he told me quite flatly when I was a b—— f——, which was most of the time, and he suffered my mistakes with good humour. Also, I realize now that he looked after me both morally and physically much more than I knew at the time. He was thirty-two years old and I was nineteen when I arrived, so it was natural for me to amuse myself with younger men. Besides, he was engaged to be married to Sally Major, who was obviously a much more interesting companion for him than I.

I had arranged to go home for a winter's holiday in 1914 after his wedding in the fall, so we had to get a house built for him that summer. The preparations for the house we did ourselves, the first job being to dig a cellar the size of the proposed building. This we did chiefly with a road-scraper, which was a sort of large, glorified shovel, hauled by two horses. The only piece of that cellar hole which I threw out by hand, was the last piece, upon which the horses used to scramble out of the hole with a scraper full of earth.

The cellar was an important feature of a Canadian house, as it contained the furnace for the central heating, and was the only frost-proof store for potatoes and similar goods which would be spoiled by getting frozen. Incidentally, onions so spoil. I saw a case of them in the store at Barloe one day, and visions of fried onions came to my mind so strongly that I purchased several pounds and placed them in the sleigh. I stopped at Henderson's for supper on the way home, and left my parcels outside in the sleigh for some hours. It was forty below, and when I got home and took out the onions they were frozen right through like balls of glass, and when I thawed them out they were useless.

The walls of the cellar were of brick, and formed the foundations for the house above, which was built entirely of wood save for the chimneys. A jobbing bricklayer and carpenter from Beaver Lake built the house, while George hauled the lumber, and I did the farm work. We had no bed for this craftsman, but he seemed quite happy on some horse blankets in a corner of the shanty.

Canadians, as a rule, did not build any open fireplaces in their houses, relying on the central heating for warmth, and running a stove-pipe from the cooking stove into the chimney. But the English settler with memories of England's homes had always one fireplace. He did not light this fire with the idea of warming himself, but it seemed like home to watch the flames flickering up the chimney. The Canadians laughed at this fancy, but to me it was very understandable. Some of my most pleasant memories are of Henderson's fireplace. As he said, to be able to tap one's pipe out in comfort, and throw nut shells into it, made it well worth the outlay.

The news of the outbreak of war came in August, but very few young men enlisted immediately. We were in the middle of harvest, and a European war seemed very far away, while the needs of the land were urgent and important. The general idea was that after threshing, if this war business was not settled by then, we might consider it.

George was getting another young lad out from England to take my place, but he was not arriving until after Christmas. My idea was to go home for the winter, and see if I could get my father to finance me in a farm in Canada. I had picked out the half-section I wanted to buy, and intended to return in the spring should my father put up the capital. If he would not, I was coming back to work for Billy Page, as I had sensed that the newly-married pair would prefer a stranger as hired man.

George was married in early December, and I drove him up to the little town of Binden, a replica of Barloe and hundreds of others, for the wedding on the day before I left for England. And in the little wooden church they were married, and afterwards I drove them to the depot and saw them off.

A CANADIAN INTERLUDE

I turned away from the depot when their train was lost in the Western haze, and walked over to the cutter feeling very lonely. George and Sally Major. No! They'd vanished. Now it was Mr. and Mrs. Hartley. Damn it, I'd lost George. Like losing a leg almost. Duke nuzzled my hand. 'Hell!' I said to myself. 'I'm getting weepy like a drunk. Never mind, Duke.' I hopped into the cutter. 'Come on, you little devil. Set 'em alight.' I drove back to Sally's home, where the wedding party were making merry.

Late that night I drove back to the shanty, and the feeling of utter loneliness returned. I put Duke in the stable, and fixed him up for the night. All our other horses were turned out, and I was leaving Duke at Henderson's in the morning.

As I walked up from the stable I noticed George's new two-storey house. It shone up in the bright moonlight, dwarfing the little shanty, and making it look like a hen-house. I wondered if George would be happier, or as happy, in the house as he had been in the shanty. I went into the shanty and lit the lamp. Brrr! It was cold. I lit the fire and looked round the shanty. It was sort of bare somehow. Of course, all my things were packed in my trunk. I was going home. I sat down and lit my pipe.

I was going home, home to Partridge Farm. . . . Damn it! This was home. This dirty shanty. Here I had been happy; yes, happier than I had ever been. England would be strange. I'd feel like a foreigner. God! How lonely I was. George was gone, not just for a week but for ever. He would come back with his wife to the new house under the bluff, and probably put pigs in the shanty. Oh, hell!

175

A CANADIAN INTERLUDE

I blew out the lamp and got into bed.

I wondered how George was making out as a married man. Funny, I had slept with him for three years, and now he was sleeping with Sally Major. Nice girl, Sally. I liked her eyes and the tilt of her chin. She was awful fond of George, too. There must be something to this love business. Still, of course, George was well fixed. He had a house and three quarters of a section of land. Would she have been so keen to marry George if he had had nothing?

Still, that was George's worry, not mine. I was going to England. I wondered how it would all have panned out when I returned. Perhaps I wouldn't come back. This blinking war, now. Not going too well. Oh, why the hell had George got married? Why was there a war? Why on earth was I going home? I fell asleep.

I woke next morning about seven, and was glad to find my fire still in. I pulled on a pair of overalls, over-shoes, and fur coat, went outside and filled a bucket with snow, which I placed on the stove. I slopped down to the stable to feed and groom Duke. When I returned the snow was melted. I poured the water into a saucepan and began to dress. After a comfortable shave, I finished dressing, and got my trunk outside into the cutter. Where was my handbag? I had not seen it for years. I hauled it out from under the bed, and opened it. 'Good Lord,' I murmured. 'That's the 'jamas I arrived in three years ago. Lucky I found 'em, else I'd have to buy some.' All my others had been worn out long ago as winter underwear. I put a few odds and ends into my bag, glanced round the shanty, went out, and locked the door for the first time since the shanty had been built.

A CANADIAN INTERLUDE

Joe Haines and a man named Ben Wyatt had taken up a half-section of prairie that summer and had broken nearly two hundred acres. They were going to Winnipeg for a short holiday, and were travelling down that far with me. We all met at Henderson's for breakfast, as he was driving us up to Barloe. The conversation at breakfast was chiefly about the war. The latest reports were serious. Apparently it wasn't just a job for the Old Country to put straight. A lot of men would be needed from the colonies. Fellows were enlisting all over Canada.

'Well, we're foot-loose,' said Ben. 'We'll study this recruiting business in the Peg. What about it, Pop? S'posing we don't come back. I sorta feel we got to look at this war a bit serious. Will you crop our place on shares if we do?'

'Don't you be fools. That's a nice half-section you and Ben got. The war'll be over long before they'll make soldiers outa you. France is a hell of a long way away. Somebody got to farm.'

'But will you do it, if we are fools, Pop?' asked Haines seriously. 'You got all our papers here, and the key of the shanty. We bought her for ten dollars per, and paid two down. Gotta payment of four hundred to make next fall. I somehow reckon we'll take a whirl at the army, eh Ben?'

Ben nodded. 'There's a heap of fellows going, Pop. Jack Cooper went last month. Old Gordon's gone, and three of the Macdonald boys went last week.'

'But there's no time to get an agreement drawn up.'

'Agreement, hell. You crop her. Get old George to help you. Good enough, Joe?'

177

'Suits me,' said Haines. 'It's a bet, Pop?'

'Oh, all right,' said Henderson, 'anyhow, you'll be back before seeding. Betcha you're back next week.'

I listened to this conversation rather guiltily. I discovered that I had been so happy and interested in Canada, that I had lost all interest in England and England's troubles. Yet here were Haines and Wyatt, who had all their hard-earned savings invested in their farm, seriously considering that they ought to enlist. Dash it! It was time that I remembered that my home and people were involved in this war. Well, I was going home anyway, so I would find out the real state of things.

'What about you, Jim? Coming back next spring, or are you going soldiering?' asked Henderson.

'Oh, if the war's over, I guess I shall be back in time for seeding.'

'Like hell you will,' said Haines. 'Think your father's going to finance you in a farm over here when he's got one at home for you? 'Sides, you'd be a blame fool to come back. Betcha I'd like the chance to farm at home. No work to do. Just ride round and do the heavy. I guess it'll be hard luck on you if this war ain't soon over, 'cause of course, you'll join up.'

I supposed that I would join up, but I was more immediately concerned with leaving a place where I had been so perfectly happy and contented for three years. Still, good-byes to Mrs. Henderson and her kiddies were at last said, Henderson drove us to Barloe for the last time, and when the train pulled out, I confess it, my eyes were wet with tears, and they are not exactly dry at this moment as I write.

A CANADIAN INTERLUDE

I said further good-byes to Ben and Joe in Winnipeg, and journeyed down to New York, as I was sailing on the *Lusitania* from that port.

It is curious how little trivial things stick in one's mind to the exclusion of more important ones. All I can remember about that land journey is that the moment we crossed the Canadian line I seemed conscious that I was no longer under the British flag. I stayed one night in Buffalo, which was full of Germans and had a definitely German bias at that date. However, I made friends with the Scottish bar-tender at the hotel, and he shepherded my steps that evening.

Of New York I remember two things, one being that the taxi-driver, who drove me to the Belmont Hotel, refused to take Canadian money, and was most offensive. The policeman who was called in to settle the dispute treated me in a similar manner, and I was at the age to resent such treatment, and did so.

After much bad language the difficulty was solved by my getting some Canadian bills changed into U.S. currency at the Belmont, an hotel which catered for Canadians. The only other thing I remember was that I ate a good meal of devilled kidneys in the grill-room, for which they charged me four dollars, and the waiters watched my prowess at the table with amazement and admiration.

We saw a few destroyers on the voyage, and I heard a lot about the war, which made me realize that it was a serious business. I also realized afterwards what a ghastly business the sinking of the *Lusitania* must have been.

I forget which land we first sighted. It may have been

Ireland, or it may have been Wales. But it was green. That seemed wonderful, for I had left a white world. I do not think I shall ever forget that first sight of British coasts.

England, when I landed, seemed strange and crowded. There were people everywhere, and I felt an alien in my fur coat. And if Liverpool seemed crowded, London was a black dirty hive. There were soldiers everywhere, and all the talk was of war. I taxied across London to Waterloo station where a girl presented me with a white feather. This England seemed a very strange country, and I preferred Canada.

But when, on my arrival at my home station, I could see Tommy and the trap outside, and found my father beaming on the platform, I knew that I was home at last.

PART III
THE WANING OF THE GLORY

BOOK FOUR

THE WAR AND THE SITE CLOSES

CHAPTER XVI

In spite of the war, I found home life little changed. My father was decidedly more crippled with rheumatism, and walked with two sticks, but mentally he was as alert and vigorous as of old, while my mother did not appear to have altered at all. She was still a staunch churchwoman, and took me to service with her on the Sunday morning after my arrival home.

That also seemed unchanged save that the congregation sang 'God Save the King' with martial ardour rather than reverence at the close of the service. When the clergyman came to the finish of the exhortation with the words 'Saying after me', there was the same old 'rumble rumble rumble' as we knelt down. The choir still sang the fugue at the end of the 'Te Deum', which gave me great joy. The same choirman boomed out the bass bit of it: 'Let-me-nev-er-be-e-e-con-fow-ow-ounded'; and the congregation heaved the same unconscious sigh of relief when the organist successfully brought his wandering vocal flock into the fold on the final 'confounded'. The schoolchildren made just the same clatter when they came in, and presumably the same two sparrows twittered in the roof.

THE WANING OF THE GLORY

I renewed my friendships with the farm men. There were a few new faces, and one or two old friends, notably Dick Turpin, had died, but the majority were just the same. The foreman 'allowed as 'ow I'd fallen abroad smartish'. The shepherd asked: 'Did 'ee 'ave any sheep out abroad?' and when I told him 'No', he shook his head as though Canada must be a poorish sort of place. The groom-gardener welcomed me with open arms. 'I've a got a stunnin' trip o' young ferrets, just fit fer work, and there's plenty o' rabbits. The Guvnor han't bin well enough to go atter 'em, and they keepers don't half do it.' The old dairyman said: 'Guvnor'll be main glad to 'ave 'ee back. 'Ee do falter a bit. You be gwaine to bide yer, I s'pose?'

The dairyman at the home dairy surveyed me with scornful approval. 'Humph,' he snorted. 'Made zummat out o' 'ee abroad then, somebody. 'Ave 'em taught 'ee any sense?' I grinned and said that they had done their best. 'Then let's hope thee't use it. There's a duty in front ov 'ee.'

'Soldiering?' I queried. 'No!' he said shortly, and continued his churn-washing, as though the conversation was finished. 'Then what duty, Frank?' I asked. 'If thee cassent see it, 'tis no use talking. Why, damn it altogether, 'tis stickin' out a voot.'

I knew what he meant, and about a fortnight after my return we came to it. During this time I had got some idea of the war position as related to our immediate neighbourhood. Nearly all the lads I knew had enlisted, and I decided that I must follow suit. Quite frankly, I don't think that I came to this decision on any patriotic grounds. I had been brought up during a period when soldiers as a class were looked down on, and I did not want to become one. I wanted to return to Canada, but at the time to enlist was the thing to do, and youth usually chooses the popular thing as the easiest way out. Besides, at my age I did not want to be out of the hunt.

As the war had completely prevented any idea of a holiday in England, my father asked me what I was going to do. I told him that I had intended returning to Canada in the spring, but that under present conditions I must enlist. When I said this I suddenly realized how old and feeble he was. 'I can say nothing against your wanting to enlist,' he said heavily, 'although God knows I want you here. I never thought to see a son of mine a soldier, but, at your age I should enlist as things are to-day.'

Thinking that I would prefer to ride, I tried to join the Yeomanry, and, to my great annoyance and astonishment, I was rejected as unfit on the grounds of flat feet. This was a great blow to my pride, as I had always reckoned that I could hold my own with other young men at anything until then. My feet may not have been of the regulation pattern, but, dash it, I could ride. I returned home railing against the authorities, who did not know a good man when they saw one, and did not realize my good fortune in the least.

My father scarcely disguised his satisfaction at the way things had turned out. 'Now, don't think about Canada, Jim. There's a man's job here for you, and I need you. 'Tis getting beyond me. I can't do without you.'

This was an appeal which I could not ignore. There

is, and I suppose always will be, something in family ties. For a strong independent character like my father to admit that he needed me, his son, touched me very much, and it must have cost him something to say. We fixed it up that I should help him at home until the war was over, when the idea of my farming on my own in England or in Canada should be reasonably considered. So I wrote to Billy Page that I should not be returning until this German squabble was finished, and settled down in the routine of English farming once again.

And it was being carried on in just the same manner as of old, with the Hampshire Down flock as its ruling motive, against which the dairying side was rebelling as before, but much more strongly. Naturally, I considered many of the methods, especially the methods of cultivation, as being archaic and expensive, and having now the authority of experience, I pointed out this to my father and the foreman. Those two worthies had seen a heap of young men make mistakes through scorning old and tried things, and although they listened to me with interest, they altered nothing.

At this time wages were still at the pre-war figure, and only a few of the younger labourers had enlisted. In the spring of 1915 the military authority commenced to build camps all around our neighbourhood. This altered the labour market with a vengeance. Boys who were not yet of military age, were at a premium, and could get three times as much money as their fathers, by working for the contractors who were constructing the new camps. At that period the farms in our neighbourhood would have been denuded of almost all labour, but for the fact that the married men lived in the farm cottages.

THE WANING OF THE GLORY

The labourers' wages rose from twelve shillings weekly to eighteen shillings, with the customary effect of rendering the men less valuable as workmen. I do not say this in any sarcastic spirit. I am simply stating a fact. When an employee thinks that he is indispensable to his employer, his value is lessened, and this sudden rise in wages of fifty per cent told all the men that they could not be done without. Young carters being especially in demand for camp-hauling, we were hard put to it to man our six two-horse single-furrow plough teams, so I suggested a carter driving three horses on a double-furrow plough.

The foreman said flatly: 'I tell 'ee double ploughs bain't no good on thease farm.' I said that I knew that they would do; whereupon he said: 'I do know. Nobody can't tell I nothin' about a plough.' 'No!' I said. 'That's your trouble. You don't know, and you can't be told.'

There we were in six months back at the same deadlock, as when I left home nearly five years before. However, the shortage of labour forced my father to give a double plough a trial after the harvest. He left the choice of make to me, and I got one with rolling coulters in addition to those of the usual knife pattern.

When it arrived, I asked the foreman if he wanted me to start it, and was informed that he had driven a double plough some thirty years before. So I left them to it. They started with it on the piece of land from which the mangolds had been carted. This land was covered with mangold leaves, which were pushed up by the knife coulters into a tight heap in under the beams of the plough, thus bunging it up solid. I arrived in the middle of this, and the foreman said triumphantly: 'What did

I tell 'ee?' I inquired the whereabouts of the rolling coulters. 'Oh, they wheel things? They bain't no good.' 'How do you know?' I asked. 'Have you ever seen any before?' He was forced to admit that he had not, whereupon the carter chuckled.

I am afraid that I appear rather priggish over this business, but I am telling it as it happened. I will readily admit that the foreman was a good, reliable, honest servant, but he could be, and had been, very annoying, and I could not hide my satisfaction at getting one up.

I sent for the rolling coulters, took off the knife ones, and fitted the new type. 'Now then, Bill,' I said to the carter, 'get on with it.' The horses started and the plough went smoothly along, the rolling coulters dividing the mangold leaves, and doing their job perfectly. The carter's comment, 'Coo, I'm b——' speaks for itself. And later on in the day I overheard him say to his mates: 'There be zummat in wot 'ee says.' That was indeed high praise.

From that date more and more responsibility devolved on to my shoulders, and I began to forget Canada. One transplants easily in youth. I can see now that I was then obviously the right man in the right place. Until that date English farming had always been equipped with a sufficient staff of labour to cope with any weather contingency. Then we had only a bare sufficiency to carry on that system of farming under ideal weather conditions, which never continued for long. Consequently, the experience of a man who was accustomed to farming with almost an entire absence of labour, was, to say the least of it, useful.

And I could do any and every job myself. That beats

employees every time. I can remember fitting sheaf-carriers to the binders for the next harvest. These appliances drop the sheaves in fours or fives at regular intervals, thus saving the labour of gathering the single sheaves together from where they have fallen from the machine. The men said that this hindered them in 'hiling' the sheaves. I laughed, and asked them how many acres a man could 'hile' under the old method. They put it at six acres daily, so single-handed I 'hiled' sixteen acres of oats one day behind a sheaf-carrier, after which I heard no more about its disadvantages.

I think that my father held the balance between 'impetuous youth' and 'crabbed age' very skilfully and justly, playing us off against each other to his own advantage. At this sort of thing he was a master. He paid me one pound weekly in addition to my board and lodging at home, and this cash payment rose to thirty shillings at the end of my first year at home. And he was a man who always reckoned to get value for money.

Of course, every product and every need of the farm mounted steadily in price as time went on, and it was this fact which led me to my first business gamble on my own account. Egyptian cotton cake, which had remained stationary at about five pounds per ton for several years, now had reached ten pounds per ton, at which price a merchant in the local market town besought my father to purchase for forward delivery. He refused, and rather guiltily I bought twenty tons on my own one September for delivery after Christmas. In the following January the dairy and flock of sheep required large quantities of this feeding stuff. My father found out in the market that the lowest current price of this commodity was twelve pounds per ton, whereupon I offered him my purchase at eleven pounds ten shillings. He was annoyed in one sense, but greatly pleased that I had started in business, and paid up smilingly, saying: 'You've made some money, Jim, but you've learnt nothing. Go on until you lose some, and then you'll have learnt a valuable lesson. I'll finance your operations.'

In the fall of 1916 my father sold off his Hampshire Down flock. Our regular staff had so dwindled that the needs of the flock had become a constant nightmare. Besides, sheep were dearer than they had ever been in his memory, which made it seem sound policy to sell out. The shepherd was heartbroken, although we were only giving up the breeding of sheep. We were still going to keep a fatting flock going, by buying in store lambs according to the season's supply of keep, and fatting them for the local market. But our shepherd had always had a regular breeding flock, and could not entertain an existence without them. 'I don't want to leff, zur, but I shan't be 'appy wi'out a lambin' flock.'

Owing to the shortage of labour he had no difficulty in getting another post in charge of a breeding flock, and in his place we acquired another shepherd to look after the proposed fatting flock. This shepherd was a man of parts. As my father said: 'He had an open mind.' He certainly needed one, for sometimes he would be swamped with sheep, and at others there would perhaps be none at all for several weeks until a favourable opportunity to buy store lambs occurred.

This type of sheep farming suited us all during that difficult time. It suited my father, as he was now master of the sheep situation instead of the flock ruling his farming operations: it suited the foreman as there was less urgency in the necessary work for the sheep; and it suited me, as it enabled me to start trading in sheep on my own account.

I began by buying little lots of store sheep at the local markets, and paying my father for their keep until I sold them. As sheep kept steadily going up and up, I kept on making money at this game, and thought it was a simple business and also that I was a clever chap. My father egged me on.

Accordingly, my ventures got bigger and bigger, and one June I bought a hundred regular draft ewes from a neighbouring farmer for three guineas each. I paid my father sixpence per head per week for their keep, and a month afterwards I offered them for sale at a local fair The bidding only reached sixty shillings a head, so I brought them home again, in spite of my father's 'First loss is best loss, Jim'.

I tried to sell them again at another fair about a month later with a similar result, and eventually disposed of them at sixty-two shillings per head some three months after I had purchased them. During this time my father entered up against my name his bill of fifty shillings for their keep each week with great glee. This deal cost me nearly forty pounds, which loss he said was buying experience well worth the money. I do not think that I profited by that experience as I should have done, although it steadied my immediate trading operations.

I was becoming more responsible for the management of the working of the farm, but my father, wise man, kept the business end under his own control, and he would hobble about the markets and fairs on his two sticks, thoroughly enjoying the combat of buying and selling. Only when his rheumatism prevented him from travelling was I entrusted with any business.

The first time this happened I was sent with the dairyman to buy three heifers and calves at the local market. Full of pride, I set off with him in the milk float, but he soon pointed out not only the impossibility of satisfying my father's requirements, but also my own incompetence for any business of this character.

'Now, 'tis no manner o' use fer you to think as 'ow you be gwaine to be clever, and buy zummat cheap. There's a limb ov a lot o' cleverer vellers than thee in market. 'Sides, whatever we do buy, 'll sure to be too dear fer the Guvnor. Best thing fer we to do is to buy dree good uns, and then all as 'ee can grumble about 'll be the price.'

The more I thought this over the sounder it seemed, so we picked out three of the best heifers, and bought them at about forty pounds each. When we returned with our purchases my father hobbled over into the yard to inspect them. He grudgingly admitted that they seemed to be 'niceish' cattle, but when I told him the price he raised his two sticks to high heaven, and said: 'My God! The best cow I ever had I gave eight pounds for.' And he hobbled away leaving me with the impression that bankruptcy was imminent.

I must have shown my dismay, for the dairyman said: 'Now, doan't 'ee worry. We've a got dree good heifers, and 'ee do know it. You didn't expect fer un to praise 'ee, surely. 'Twouldn' be natural. Praise do make young folk uppish.'

THE WANING OF THE GLORY

Very certainly I was never allowed to get uppish. Looking back on my youth I cannot imagine anyone being very uppish with my father, although he was distinctly uppish with most people.

Although my father was making money out of his farming at that time, neither he nor the men had any joy in it. There was no definite system. If the labour shortage made it impossible to do a certain job at a certain season, the fact of its being left undone was resented strongly. For instance, to have a good plant of rape and turnips, and to be unable to get them hand-hoed and singled out, rankled. Thinning them out by repeated harrowings was reckoned a poor job. To see banks untrimmed, ditches uncleaned, field corners not dug, and one's farm generally untidy, hurt one's proper pride. But there it was. There was only barely enough labour for essentials, and the frills had to be cut out. From 1916 the farms in the countryside were allowed to deteriorate or to 'go back', a local term which describes the situation admirably.

Not only did the farms deteriorate in general appearance, but, very definitely, they did so in fertility. In our district, it was possible by good farming to grow two corn crops in four years on most of the land. Food shortage caused the Government to ask for an increased grain acreage, labour shortage fostered this demand and high prices made it an attractive proposition. Accordingly, sixty per cent or even more of the arable land was put into corn each year with a consequent loss in fertility. However this loss was not apparent at that time, as the land had been treated fairly for generations previously, and could stand a fair amount of over-cropping. But even if the corn-grower did well in those days, the grass farmer and milk-producer did better. Milk went up as did grain, and there was an eager and increasing demand for it.

Home life was naturally not so spacious and carefree as in pre-war days, but there was no lack of prosperity. Amusements and pleasures were practically non-existent. One farmed, one made money, but war conditions made it impossible to enjoy it, so one carried on, as did every other class, hoping and longing for the finish of the war, and a return to normal conditions.

It may seem curious, but I can remember less detail of that war period of farming than of earlier days. That time seems now as an unpleasant dream. There was no planning for future years—the future of all England seemed very dark and uncertain—but there was always more necessary work in front of us each morning than could possibly be done in the day. So we did what we could, day by day, and left the future to Providence.

As livestock grew scarcer and dearer, it became necessary to go farther afield to buy store sheep for the fatting flock. Our bold Tommy, who was still hauling the governess cart as of yore, proved unequal to these new demands, more especially as the road traffic—ours was a camp district—had increased greatly. As my father regarded motor cars as his natural enemies, it was with many misgivings that he purchased a car, and instructed me to take on the duties of chauffeur.

These duties were by no means easy. My father knew all about driving a horse and trap. Indeed, to go out with him for a drive was a liberal education for anybody. Amongst other things you learnt that, while

THE WANING OF THE GLORY

Tommy ostensibly did the work, it was necessary for the driver and all the passengers to assist. The whole journey was a series of adventures and difficulties which it was only possible to overcome by team work. You sat back when going downhill, and you sat forward when going uphill. If you did not you were told about it plainly and very forcibly. On the level parts of the road there would be other trials. You circumvented the Scylla of the oncoming motor car on the one side, to be confronted with the Charybdis of a piece of paper in the hedge, at which Tommy was expected to shy, on the other.

Amongst other things you learnt that no motorist had any road manners at all. Incidentally, my father's were atrocious. When we passed a broken-down car, Tommy would be urged to his most spirited gait, and with a conscious air of superiority we bowled by. Motor cars meeting the trap, and endeavouring to pass at speed, would be slowed down by my father carrying his long whip at the horizontal right across the oncoming car's path. On narrow roads Tommy was halted bang in the middle in a death or glory fashion, thus forcing the car to stop and allow an older and more dignified civilization to pass by at its leisure.

But as Tommy's highest speed was about eight miles per hour, and his customary gait only six, a motor car was the only way of accomplishing any journey of longer than eight miles. So, as I have said, a car was purchased, and I learnt to drive it. When I was pronounced efficient by my teacher, and had managed to return safely after a few journeys on my own, my father entrusted himself to my tender mercies. It was either that or letting me go alone to buy sheep, which would have been much too expensive a business in his opinion, and I have no doubt that he was right. So he put his own feelings on one side, and braved the terrors of the road with me.

Mind you, on these expeditions he was still captain of the ship, and he hoped master of his fate also; I was but the mere engineer. He sat with one eye on the road and the other on the speedometer. Anything over fifteen miles per hour was considered speeding, and I drove that car for several months to the accompaniment of 'Steady! Steady!' For a long time he still sat forwards going uphill, and sat back with his feet braced against the footboards when going downhill. We literally stalked our corners, and we went carefully and slowly in traffic, usually in low gear. But we never had any serious mishap, and after a while he quite liked this new mode of transport.

And then, quite suddenly, in the middle of the harvest of 1917 my father died. He was ill only a few days, and then he was gone, leaving a gap which it seemed could never be filled. I do not know whether I have drawn him quite fairly in this book. Like King John in one of A. A. Milne's rhymes, 'he had his little ways', and I seem to have written chiefly concerning these. But unlike King John, if history tells us true, he had his 'big ways', and they were splendid. In my youth I railed often at many little things he did, but in every big crisis in my life I never turned to my father in vain. That is, I think, a wholesome memory for any man to leave to his son.

I have often thought about my father since his death.

THE WANING OF THE GLORY

I suppose that he might rightly be described as one of the Victorians. He certainly possessed the virtues of that era, and possibly the vices also. Personally I have no patience with the modern habit of sneering at the Victorians. Whatever else they may or may not have done, they did their job. They worked, and worked hard. Today we are rather apt to point to them as people who lived their narrow sordid lives solely in the pursuit of material things. Bet that as it may, who are we to criticize them? The Victorians accomplished things, possibly to our modern view, in somewhat nasty fashion. Is it not possible that, by comparison, we Georgians accomplish nothing of note, and also that our efforts are conducted in a manner less admirable?

CHAPTER VII

Immediately after my father's death responsibilities were showered on me from all sides. I was a woefully thin and inadequate peg to fill such a large hole. A bare week ago my father had been there to make the final decision about anything, to act as the last court of appeal in our farming problems. And now he was dead and buried. It seemed unbelievable. Was I never again to go indoors, after my early round of the farm, and mount to his bedroom to report?

Still, I had not much time for these vain wonderings. Suddenly I had become the last court of appeal. We were busy at harvest, and decisions had to be taken. So I took them, many wrongly I expect, but I took them. I had to do so—the men saw to that. They seemed to delight in finding problems for my handling, in order to find out how I would shape.

The foreman decided that nothing was ever quite fit to stack, and presumably never would be now that my father was gone. 'If you picks that up to-day, mind, I bain't responsible.' 'No!' I said. 'You aren't, anyway, so we'll pick it up.' And I stacked everything on my own judgment, and chanced the consequences. Providence and the Clerk of the Weather proved to be on my side on this occasion, for everything went off all right.

Then, before we had finished harvest, one of the men, Bill Avery, who lived in one-half of a double cottage, having a common front door, gave me a week's notice on Friday night. There had been domestic trouble brewing between him and his neighbours for some time, and, but for the scarcity of labour, either Avery or his neighbour would have been sacked long since. But Avery was the better and more valuable man, and knew it.

'I bain't gwaine to bide long side o' wold Fray and thic 'ooman o' his no more. Either you gets rid o' they, or I gies 'ee a wik's notice.'

While he launched forth into a list of Mrs. Fray's iniquities, I thought it over. I would have a difficulty to replace Avery, but, dash it, badly as I might be managing the farm I was supposed to be managing it. Anyway, Avery was not, and very definitely he was not going to do so. To let him go was bad, but to keep him on his own terms would be worse. So I cut into his list of wrongs with: 'That's all right, Avery. I'll take your notice as from to-night as you wish. Let the foreman have the key of the cottage next Saturday morning when you leave.'

'Ho!' he snorted. 'I be to go, be I? Be I a dog to be turned out when thee's alike?'

199

'No,' I said, 'if you must know, you're a silly fool, and you're leaving your job and your house at your own wish. You've given me a week's notice, which I've accepted, and that's that. Good night. I'm too tired to argue.'

Social reformers may consider this procedure heartless, and point to it as an example of rural tyranny, but subsequent events proved that it was the right course to take, and it definitely established my position as boss with all the men. If any one can point out a better solution to the difficulty, always bearing in mind that my job was to run the farm, I shall be glad to hear of it.

This business with Avery impressed even the foreman. That worthy unbent quite graciously a day or two afterwards, and 'allowed as I'd done a good job'. He was a veritable Biblical patriarch in appearance, and held the same ideas concerning women in their relative importance to men in the scheme of things as his Old Testament prototype. 'Ay,' he said, fingering his beard, ' 'tis a good job as Avery's gwaine, but you maun finish it. Fray's missus be that puffed up wi' pride awver it, you maun set her to rights. She do fairly flaunt up and down the village. You ketch she one marnin' an' tell she off proper. It don't do fer a 'ooman to get like that.'

On reflection, this seemed to be sound advice, so I plucked up my courage and called on Mrs. Fray a few days later. She greeted me with a beaming smile, and almost bobbed a curtsey. Was I not her friend? Had I not taken her part against the hated Averys? She invited me to go inside.

This would never do. I must say what I had to say on the doorstep, where the other village dames could overhear, or the effect would be lost. 'No, Mrs. Fray,' I said. 'I won't come in. I just called to warn you to be careful of your ways. The Averys are leaving, and I shall have some more folk next door in a day or so. Now mind, if you can't agree with them, or if I hear one word of complaint, you and your husband will have to leave. I'm too busy to be worried with such nonsense, and you are old enough to know better. Good morning.' And I hurried away up the lane feeling thoroughly ashamed of myself. Verily, there was a lot more to farming in those days than the mere growing of crops or raising of stock.

However well or ill I may have managed the farm during the year following my father's death, there is no doubt but that it was managed profitably. It was impossible to lose money at farming just then. Wages had gone up, but there was still a considerable time lag between them and the selling price of agricultural produce. For instance, I think it was in that summer of 1918 that we seeded ten acres of vetches. These had been planted for sheep feed, but as our fatting flock was a variable quantity, in this instance we had not required them. They yielded twenty-four bushels per acre, and I sold them for thirty shillings per bushel. This crop realized thirty-six pounds per acre, which was at least three times the freehold value of the land. Fancy feeding stuff of that value to sheep!

As my father had died in August 1917, his executors received a year's notice to quit the farm at Michaelmas, 1918. I had nothing like enough capital to take it on, but it was arranged that I should borrow the necessary from my mother at five per cent, and have a try at it. Accordingly, I interviewed the estate agent, and was informed that I could have the farm at nearly a hundred per cent increase in rental. I was also told that the estate was not going to bargain with me, that they could in all probability obtain an even higher rent in the open market—this was true—and that I must take it or leave it.

During the last two years I had been called up by the military authorities many times, and it was about this time that I went up for another medical examination. The doctor in charge asked me how far I could walk at my own pace. I thought of my plodding behind a grain team in the Canadian snow, and told him twenty miles. He examined my feet again, informed me that I could not walk five miles, and that accordingly he should reject me for military service of any kind.

Being therefore a comparatively free man, I arranged to take the farm at the estate agent's terms as from Michaelmas 1918, and I got married that summer.

That seems a rather bald way to mention such an important thing in a man's life, but this book is intended to be primarily about agriculture. So of my wife I will say but little. She has put up with me and my varying moods for a good many years now, for which I hope I am sufficiently grateful. A husband, obviously, cannot be unbiased in his opinion of his wife, but the only serious differences which we have had up to date have been concerning the mysterious disappearance of all combs from my dressing-table since she succumbed to the prevailing fashion for short hair.

My mother refused to carry on in the farmhouse after my marriage on the grounds that it was much too big and that to live alone in it after her full and busy life there would be too painful. So she made her home with one of her sisters, who lived a few miles off, and insisted that my wife and I should live in the farmhouse. I can see now that this was a big mistake on my part. We began where our respective parents had left off, which is always wrong. I must have had some faint glimmerings of common sense even in those days, for I can remember that I wanted my mother to stay in the big house, and let my wife and me go into a little house in the village. But I allowed myself to be over-persuaded against my better judgment. I suggested letting the house, and was told that the little I should gain by doing that would be more than lost because I should then not be living in the middle of my job. So I came back from my honeymoon to the house in which I had been born, and began gathering my mother's last harvest. I was able to thresh her entire crop out of the harvest field that season, and to settle up all accounts with her at the end of her tenancy in October.

I took off the whole of her stock and implements at valuation, which was calculated by a local auctioneer. There were no sheep, as I had so arranged things that the last lot of my mother's fat sheep were sold in early October. Everything was dear, very dear, including sheep. Whilst I was quite willing to borrow money to purchase horses, cows, and implements at times prices, to borrow money to buy sheep at these high prices seemed to be too much of a gamble. Even the auctioneer, who was a sheep man, advised against such a risky course. Horses and implements were necessary; cows would bring in a return from their milk almost immediately; but it took a very long time to obtain any return from sheep. Besides, their cost would have increased my debt greatly and it was alarmingly large without them.

I often chaff my mother to-day about her wisdom in going out of business at that time, and my own corresponding foolishness. But I knew the farm, I knew the men, and I thought I knew how to manage the place. Of course it is easy to be wise after the event, but what a fool I was, what a silly, silly fool! What I should have done was to have managed the farm for my mother at a wage until to-day, for in spite of the present depression in farming, now is the time for a young man to start.

But I did not think. I was born on that farm; it had always paid handsomely, and very certainly it was a paying proposition then, so I started farming on my own account eagerly and with no fears for the future, save the all-pervading fear of the outcome of the war.

The estate had reduced the number of farm cottages considerably, so I sorted out the men in my mind, and engaged those whom I considered the most suitable. The foreman retired gracefully to a cottage of his own. He had a nice bit of money, my father had left him fifty pounds, and, as both he and his wife drew the Old Age Pension, they were fairly comfortable in their retirement.

I cannot remember that I had decided on any definite rotation to follow. Broadly speaking, the idea was to grow a rather larger acreage of corn and of clover hay, and to summer-fallow most of the other arable land, which, under the old rotation, would have been growing feed for the flock. I would grow a certain amount of sheep feed, and get a neighbour to feed it off for me.

THE WANING OF THE GLORY

This was much the same as the last year or so of my father's farming, the only difference being that I proposed keeping more milking cows and no sheep of my own.

As all the world knows, the war ended in the November, and it was as if a heavy weight had been lifted from the whole country. The reaction to this was that the whole population went pleasure mad. All classes indulged in a feverish orgy of all those sports and pastimes which had been impossible for four long weary years.

Rural communities were no exception. Hunting, shooting, fishing, and the like, suddenly reappeared in our midst. In the summer tennis parties became the order of the day. Farmer's Glory was going to be as splendid as of old, only more so. More so because we all had money to burn. I find this hard to write. It is not a pleasant thing to set down on paper what a tawdry life one lived in those few years immediately after the war. But the majority of farmers took no thought for the morrow, their only idea was to have a good time. Instead of living for one's farm, the only desire was to get away from it and pursue pleasure elsewhere.

And I was as bad, or as daft, or, possibly more truthfully, as criminally extravagant as any one. I kept two hunters, one for myself and one for my wife; and glorious days we had together with the local pack. I went shooting at least two days a week during the winter. We went to tennis parties nearly every fine afternoon in the summer, and, in our turn, entertained up to as many as twenty guests on our own tennis-court, and usually to supper afterwards. It seems incredible now, but we began—God forgive me—to have late dinner in the

205

evenings, and in addition to all these rural pleasures we journeyed far afield in search of excitement.

In short, farmers swanked. It would not have mattered so much if they had confined their swanking to their immediate surroundings, but this was the last thing they wished to do. For this, I blame the motor car. Perhaps blame is hardly the word, but it was the motor car which made a lot of this swanking possible. Its advent marked an epoch in rural affairs rather more definitely, I think, than in urban ones.

Prior to its arrival in country districts, the radius of a farmer's social activities was restricted by the capacity of his horse and trap as a means of transport, but the car infinitely extended the possibilities. Farmers now went away from home for frequent holidays, to seaside resorts and to London. They discarded the breeches and gaiters of their ancestors for plus-fours of immaculate cut, incredible design, and magnificent bagginess, in which garb they were to be found on every golf course. The blue lounge suit for evening wear gave place to the dinner-jacket.

Personally, I started golf in 1919, and in 1921 I was the proud possessor of a handicap of eight, which statement tells only too plainly the amount of time I must have spent at the game. The younger generation played tennis still more seriously. Instead of being content with local tennis, which was then, and is now, of a fair standard, they joined clubs in towns thirty miles off, where they were taught the correct style by professional teachers. It is curious to note that in spite of this, it was usually the middle-aged folk who won the local tournaments against these exponents of style.

THE WANING OF THE GLORY

I imagine—no, I am sure—that the foregoing will annoy a great many people, but I have tried to write it truthfully, and I readily admit that I was one of the worst culprits in this feverish pursuit of pleasure. Possibly, and I think probably, people engaged in other industries did much the same things at this date. Anyway, in the farmer's case, he could well afford it. I did all these things, paid the interest on the borrowed capital, and paid off about five hundred pounds per annum for a year or so immediately after the war.

My world went very well then. I was newly married, and my wife was an ideal playmate. The war was over, for ever and ever, and farming had returned to its old splendour. Farmer's Glory was then a glory of great brilliance. How were farmers to know that it was but the last dazzling flicker before the fusing?

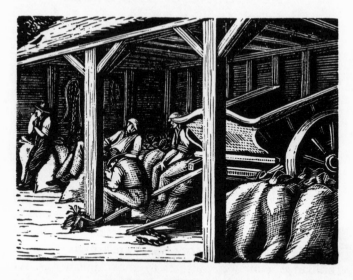

CHAPTER XVIII

And what of the agricultural labourer amidst all this prosperity and extravagance? He shared in it to some extent certainly. His pre-war wage of twelve shillings per week rose year by year, reaching its highest point of forty-six shillings for a week of forty-nine and a half hours in 1921, thereby overtaking the rise in the price of agriculture produce for the first time. His wage not only overtook the price of produce at this juncture, but it passed it.

This rise in wages was much greater than appears at first sight, as the wage then fixed was for a definite number of hours, all those hours worked in excess of this legal quota being paid for at the overtime rate. Generally speaking, the hours worked in pre-war years were about fifty-eight per week.

This fixing of wages went hand in hand with a guaranteed price for grain until and including the harvest of 1920. This guaranteed price was altered in character for the harvest of 1921, giving instead a bonus in cash per acre on land on which wheat and oats were produced. In 1921 this bonus amounted to three pounds per acre for wheat and four pounds per acre for oats. After these payments in 1921 that section of the law was repealed, leaving the part relating to wages still operative.

And here lies the farmer's chief, and I think only legitimate grievance. Control of production costs allied to a guaranteed price for grain, he tolerated, in spite of his inherent objection to any form of governmental interference. A cancellation of all control hew ould have welcomed at that time, although subsequent events have proved that this would have found him in a difficult position to-day. To control his production costs, and to leave the price of his products to chance, was unjust, but it suited the politicians' book, and no other consideration came into the question.

In my own case, I was in a mess of my own making. I should have foreseen that these good times must come to an end some time, and should have made provision to meet it. But I did not, and neither did the majority of farmers who started farming when I did. The farmer who had been farming in 1914 and was still farming in 1921 had at least some war profits to lose. But the men who started in 1920 and 1921, relying on the good faith of the politicians, were let down badly. Their capital simply melted away. And these men were for the most part returned soldiers.

Take the case of any farm which changed hands at

THE WANING OF THE GLORY

Michaelmas 1920, and again, say, in 1927. At the first change the incoming tenant would have been forced to stock the farm on the basis of the prices realized at the outgoing tenant's sale, cows at sixty pounds each, horses at one hundred pounds each and sometimes more, and the other stock and implements at similarly inflated prices. His rent would also have been based on the prosperous condition of farming at that date. A risky business, possibly, but there was the law, which said quite definitely that his chief selling crop, grain, should be certain of a remunerative return. Instead of gambling against weather conditions and world prices, he had only the weather to battle against. For the first time in farming history he was farming on a certainty. What mattered the enormous capital required to finance a certain winner?

And so farmers tumbled over each other in their eagerness to get farms. Poor devils! They soon had a rude awakening.

At the second change of tenancy in 1927, the incoming man would have purchased his cows at about twenty-four pounds per head, and his other stock at a similar depreciation. In addition, the landlord would have had to reduce the rent of the farm considerably, in order to get a tenant, and, in the case of a large arable farm, would have probably grassed and fenced a large portion of it as an added attraction.

It is obvious that the incoming tenant in 1927 would have only needed about forty per cent of the capital to stock the farm as compared with his predecessor, and, taking into consideration the reduced rent and other advantages, he would have been almost on an economic basis. Since 1927 even, a further depreciation has taken place, and should the same farm change hands to-day, the incoming man would need still less capital. In short, every thousand pounds invested in farming in 1920 is to-day worth only about two hundred and fifty pounds, and it is still depreciating, more especially on arable farms.

I suppose that this depreciation will eventually reach such a point that owing to the deflation of capital, farming will automatically recover. Of course, politicians of every shade of political opinion are well aware of this, and are deliberately delaying any real political action with regard to agricultural matters. The fact that nearly every farmer who went into a farm soon after the war will have lost three parts of his capital does not worry them one bit. When it is fairly obvious that farming has started on the up grade once again, the party then in office will pass some agricultural legislation and not before. And it will not matter much what they pass (compulsory crossing of carrier-pigeons with parrots, in order that the offspring may transmit verbal messages, will be about the mark). If they can only time their action at the right moment, they will go down in history as the political party which saved British agriculture.

Not only did the tenant farmer suffer through the repeal of the Corn Production Act, but those who purchased their farms at that date experienced a still greater loss. Farms, during the years immediately after the war, were at a premium. In addition to the farming population, who were keen buyers, there was an influx of town folk into this market. Town traders, who had amassed comfortable fortunes during the war, proceeded to retire from business, and to settle down on farms either by purchase or rent, especially in the south of England. Their arguments for so doing were perfectly reasonable. They were prepared to pay a little more for a farm than a farmer. What did it matter if they paid a hundred pounds per annum more? They would get a nice house, some good sport, and a background for their families. Even if their farming operations cost them a little each year, what matter? They had to live somewhere.

So this class of buyer and renter came into competition with the ordinary farmer, and up went the price of farms in consequence. A tenant farmer might have rented his farm for thirty years. It was then not only his business but a home containing the memories of a lifetime. What was he to do if he gave up farming? His landlord was forced to sell, and these new buyers made the price almost prohibitive. Still, there was the legal guarantee of corn prices, and on the strength of this farmers purchased their farms or rented them at times prices.

One man said to me in conversation the other day, when we were discussing this matter: 'Grievance? I reckon I've got a just grievance. On the strength of the Corn Production Act I bought my farm and paid cash for it. I've been let down.'

I murmured that he was fortunate in that he was able to pay cash, and asked what of the man who had purchased on mortgage.

'God help him,' he replied. 'Look at old So-and-so. He bought his place for sixteen thousand pounds. 'Twas either that or turn out. He put down half of it, and the bank advanced the rest. Now it's just been revalued at eight thousand pounds, and the bank are still willing to advance half of it as before, but they want him to find a matter of about four thousand to put things straight.'

'What'll happen?' I asked.

'What can happen?' he retorted. 'Nothing! If they insist they'll bust him, and that'll do 'em no good. They'll wait for something to turn up, and meantime keep on worrying him.'

And that was the farmers' general attitude to the position. Like Mr. Micawber, we waited for something to turn up. Possibly we reduced our private expenditure in some minor way, or reduced the number of employees a little—a favourite method of doing the latter being to purchase a tractor for ploughing and threshing, thereby dispensing with some horses and carters.

It was in 1921 that the first suspicion of anything wrong in the farming world was noted. In spite of the bonus paid under the Corn Protection Act of that year, money was a little tight. The harvest of 1922 left no doubt in my mind that things were going—no, had gone— wrong. It was a wet summer, and the price of both corn and milk had slumped badly. I found that I had great difficulty to pay my interest and meet my other liabilities.

Something must be done, and that urgently.

Now when a farmer finds that his yearly sales do not bring in enough money for him to pay his way, his most usual course is to try to increase his production. In my own case, I tried to grow more corn. To-day I am well aware that this procedure was idiotic, but at that time I was not alone in this idea. I had about four hundred acres of arable land, and, in good farming practice, I should have grown two hundred acres of corn each year.

As a matter of fact the farm had been growing about two hundred and twenty-five acres of corn annually for several years. Still, I thought that I would grow more, perhaps up to three hundred acres, with the help of artificial fertilizers, which were at that time low in price compared with other commodities.

Now it is generally supposed that corn can be grown successfully year after year on the same land, so long as the requisite plant foods are put into the ground by means of artificial manures. Indeed, this is no longer a supposition, but a fact, as it has been proved by experiments at Agricultural Colleges and Experimental Farms. I am willing to admit that this may be possible in some small portions of our country, but I say definitely, from bitter experience, that it cannot be done for long on our south country hills. It may be said that I did not know how to do it, but I have seena a good many others fail at it in this neighbourhood besides myself, and I have yet to see it done with success.

As the older labourers put it: 'That chemical tackle ain't got no proof in it.' By 'proof' they mean an intangible something, which can best be described as 'essential guts'. Anyway, something is lacking, call it what you will.

What happens is this. You get good crops for a year or two. Then, for another year or so you get crops which look good, but their yield of grain in comparison with the quantity of straw is disappointing. Finally, you get crops, especially of wheat, which look well in the spring and early summer, but after coming into ear they go what is described locally as 'kneesick'. By this I mean that the 'knees' or knots in the stems are too weak to carry the weight of the ear, and a large portion of the crop 'bruckles' over, and does not mature at all. 'Bruckles' is I think expressive enough in itself to require no explanation.

It is obvious that these happenings brought me up against the law of diminishing returns year by year, against which increased grain acreage not only made no headway, but rather accentuated the discrepancy.

By 1925 the financial position was serious. Ruin gibbered in the background not so very far away. I was not a moneyed man as were many of my neighbours. Something had to be done. But what? This corn-growing business was hopeless. I turned my attention to milk.

Now this farm product had not slumped so badly as had corn. Milk products may be imported easily, but fresh milk is a different matter. The new idea of rationing and recording milch cattle seemed to me to have possibilities. Accordingly I went in more extensively for this branch of farming, and tried some sugar beet in order to give my impoverished arable land a rest from corn-growing.

This new idea with regard to milk production was that a cow was a milk factory. If you fed her with the necessary ingredients to produce milk, in addition to a maintenance ration to keep her alive, there was apparently no limit to her capabilities. Roughly, twenty pounds of hay daily or its equivalent was ample for maintenance, and about three pounds of certain concentrated feeding stuffs would produce a gallon of milk. The more of this latter you could get in to her, the more milk you would get out of her, like a penny in the slot machine.

THE WANING OF THE GLORY

I want to be quite fair about this matter. I believe that to-day research on these lines has improved results enormously, and also I know that many farmers are now making a success of this method. Enthusiastic advocates of this sytem will probably attack me for publishing my own experiences in this direction, and will no doubt say that I was an incompetent fool. That I was, and probably am, a fool, I will admit. No one is more cognizant of that fact to-day than I. But the results of other folk in this district were very similar to mine.

For the first year I got splendid results. I made a dairy of thirty-seven inexpensive cows average nine hundred and thirty-seven gallons per head. I can remember buying one old cow for fifteen pounds, and her yield that year was over fifteen hundred gallons. This seemed like 'money for jam', so away I went into cows, 'bald-headed'.

And the next year came the reckoning. I had two dairies at that time, one of thirty-seven head, and the other of forty-five head. The second year eight cows simply faded away, being unable to stand the strain. There were twenty-five cows which stopped breeding and had to be sold as barreners, making about half their cost as milk cattle. In addition milk prices were still on the down grade.

The sugar beet business was a little better, but there was not enough in it to be worth while. It was only better in that the sale of the crop brought in a little money, while the fertility of the land was being improved instead of being impaired as by corn-growing. In the first place, the railway carriage to the nearest factory far away in Norfolk swallowed up too large a

216

proportion of the price, and, in the second place, the land in our district was not well suited to this crop.

Our land was much too shallow, thus causing the beets to 'fork'. For the benefit of the townsman I must explain that a sugar beet should grow straight down like a parsnip. In my case they could only go straight down for four or five inches, and then were forced to spread out into three or four roots, or, as I have said, to 'fork'. When lifting time came, I had forty acres of things like a fanged molar tooth to extract. We extracted them, but it was a painful and expensive process, the only gas used being bad language.

Still, I do not regret growing sugar beet for two years, in spite of these handicaps, and think that on suitable land near a factory it may be a possible crop for English farming in years to come.

I can see now that this hunting about for a way out of my financial troubles was an expensive business. Farming only pays when one farms on a definite system year after year, in a rut, so to speak, from which there is no escape. This dodging about from one thing to another was expensive. Still, with corn-growing a hopeless proposition, this painful process of finding an alternative system had to be undergone. And many others besides myself plodded on, searching anywhere and everywhere for a way of escape. Mostly we failed, but in looking back on that unhappy period, I take some small comfort in the saying that the only real failure in life is in giving up. In spite of our former foolishness, let it stand to our credit in life's balance-sheet that at least we tried, and tried hard.

CHAPTER XIX

I have been informed by a friendly critic, who has watched with interest my amateurish struggles in attempting this book, that this portion of it lacks charm as compared with the earlier pages. I am afraid that this is only too true. Success and money-making, although they are often sordid, can be charming, but failure and loss can surely never be. I think that it was the placid unchanging well-being of years ago, which made rural life so charming. Possibly it is simply that age has a mellowing effect, and that one remembers only the pleasant portions of that bygone period.

I have written of the wealthy townsman, who went in for farming immediately after the war. Many of his reactions to rural life were certainly interesting. A few of these newcomers to the countryside settled down to farming, and became acclimatized very quickly, but a large proportion of them fled the countryside when farming began to be a serious drain on their pockets. One ex-manufacturer I know loved every moment of his rural existence. He was a peppery, explosive individual, who had run his works ruthlessly as a cold-blooded business proposition. To find that he was not finished with his farm employees when they ceased work for the day, pleased him greatly. To discover that his farming business was largely in the hands of the weather, partly in the hands of his employees, and only in a very minute degree affected by himself, both annoyed and amused him. 'These fellows do exactly what they like,' he said to me one day. 'They don't take the slightest notice of my wishes. They treat me as a sort of halfwit, who means well and has to be humoured. What they want. . . .' And here he trumpeted rebellion and rank heresy for a few minutes. But all the time there was a twinkle in his eye. He knew, and so did his men.

The older type of agricultural labourer regarded these 'London' farmers with suspicion. Well-meaning as some of them were the labourer did not want to be uplifted by anyone, and resented any interference in his customary habits by these interlopers, even though it might be of great benefit to him. Also he regarded these alien folk as people who had no manners. An instance of this, which deals in some measure with high society, comes to my mind.

A certain gentleman was made a peer shortly after the war, and in order to give his new rank a proper setting, he purchased a small estate in our neighbouring country. One of his farms became vacant, and was let to a new tenant. In the new agreement the incoming man was

THE WANING OF THE GLORY

to get twelve cottages with the farm. He had another farm nearby, and, in making his arrangements, he promoted a single under-carter on his home farm to be head carter for the new venture. This man forthwith made arrangements to get married on the strength of his new job.

Now the peer and his agent had made a mistake in the number of cottages, and as Michaelmas drew near they discovered that there were only eleven available. However, Miss Mills, a spinster of sixty, lived alone in one of their other cottages in the village, and looked after her widowed father of eighty-four, who lived in another cottage next door. Accordingly the agent was instructed to tell Miss Mills that her cottage would be required, and that she must move out into her father's house. Both cottages belonged to the estate, and as they were semi-detached, it seemed a reasonable way out of the difficulty.

The agent muddled the business, and was ordered out by the old lady. 'Haw,' said his employer, when he reported the result of his visit, 'you didn't handle that very tactfully. I had better see the woman.'

He called one afternoon, went inside without invitation, sat down on a chair without permission, neglected even to remove his hat, and baldly stated his business. The old lady was furious.

'I'll thank 'ee to go,' she said, 'at once, too.'

'But I don't think you realize who I am,' he said. 'I'm your landlord. I'm Lord Blank.'

'Lard Blank!' she screeched. 'Lard Mushroom, thee's mean. Let me tell 'ee as you be jist nothin'. I've a had many gentlemen in yer afore you, but I ain't never had

220

no one afore who hadn't got manners enough to take off his hat.'

'But my good woman,' the offending peer stuttered.

'I bain't yer good 'ooman. Thank God, I be zummat better'n that. Look at yer. I've a got better manners than thee, and I'm a heap better bred. That makes 'ee look, don't it? Who be you? Lard Mushroom! Jist a jumped-up little veller. Now you'd best goo.'

He retired discomfited, and told his agent to make the farmer some allowance for the loss of the cottage. The agent reported this to the farmer, who, greatly amused, said: 'Don't you worry. I can get her to shift.' The agent bet him half a sovereign that he would fail, and went away feeling pretty certain of winning his bet.

A day or two afterwards the farmer called on Miss Mills and asked if he could have a word with her on business. She invited him inside.

'It's no good my beating about the bush, Miss Mills. You know what I've come about. I realize that you have the stronger position, and I've come to ask you as a personal favour to me, if you can possibly let me have the cottage. You see, it's not my fault. I took the farm with twelve cottages, and young Frank Hardiman has had his banns published, reckoning that I had a cottage for him somewhere over here. It'll be sort of awkward if I can't find him one, as he's my head carter.'

'Ah!' said the old lady. 'Thee's know 'ow to get about things. Different from thic little toad of a lard as come yer to'other day. Don't 'ee worry. I'll be shifted in time fer young Frank. Atter all, thee's got to farm the place. Tell un to git married, and I got a piece o' china fer un fer a weddin' present.'

The farmer thanked her, and inquired as to what went amiss between her and her landlord. She told him at length, and when he cocked his eye inquiringly as she told him about her having better blood in her veins than the lord, she said: 'Ah, now thee dussent believe I? I be a ——, I be,' naming the ruling old aristocratic family of the neighbourhood. 'You come 'long o' me.'

He followed her upstairs to a bedroom, where she opened a small cupboard over the mantelpiece, and there, as in a shrine, reposed a top hat of ancient shape. Reverently she took it down and exhibited the name in the lining. It had belonged to her aristocratic sire, who had left it behind him on one of his youthful escapades some sixty odd years before.

Generally speaking, the older labourers longed for the times of long ago in spite of their increased prosperity. 'But times are better for the men now,' I said to a rugged old-age pensioner one day, when we were discussing rural affairs. 'Look at the wages they get.'

' 'Tidn what they gets. What do 'em earn? Why, when I wor a young man I wor worth dree ov 'em. Do make I fair voam at mouth to zee zum ov 'em fiddlin' about at their work.'

I met this same critic of modern times only last week, and he returned to the same topic. In the course of his remarks he touched on the proposed raising of the school age. 'Lot o' voolery,' he snorted. 'Whatever be 'em thinkin' about? Doan't 'em know what'll happen? If they keeps childer at school till they be sixteen, more'n half the maids'll be in trouble fore they do leave. Childer wants to be broke to work young, like colts. Work never hurt I, and I never knowed it hurt noboby.'

THE WANING OF THE GLORY

I make no comment as to whether his prophecy will be correct or not. That his remarks will be regarded as a foul slander on rural youth, I have no doubt, but I would point out that the speaker had lived for eighty years in a rural district, and therefore presumably was qualified to give his opinion. I sometimes wonder if the people in Whitehall are so well qualified.

Apart from the effects of the depression which was more and more rapidly creeping over the countryside, the actual farming itself lacked romance and charm at this date as compared with pre-war days. Agriculture was becoming mechanized. The horse was disappearing from the landscape, and giving place to the hideous tractor. I did not purchase one of these implements until our old portable steam-engine came to the point in its long and dignified career when the cost of adequate repairs was prohibitive.

When the tractor arrived the men viewed it with scorn. Its size as compared with the huge bulk of the steam-engine made it appear as a toy. We arranged to thresh with it a day or two after its arrival, and the feeder of the threshing machine said 'as 'ow 'ee wor gwaine to feed zo as to bring a little pooper like that up all standin'.'

Next morning, when all was ready, I let in the friction clutch on the pulley wheel of the tractor, and when the hum of the threshing machine had reached its correct note, I waved my arm to the feeder to begin. He did so, and the rich hum of the thresher died away as the machine slowed down. The men's faces were triumphant, whilst mine was the reverse. However, I noticed that although the machine slowed down, the tractor main-

223

tained its speed, and therefore the clutch must have been slipping. I stopped the engine, studied the instruction book which had come with it, and tightened various nuts and springs on the clutch.

We started again, and despite the feeder's almost superhuman efforts the tractor drove the thresher so fast that he was nearly shaken off his perch, and was forced to admit defeat.

But all the time, at the back of everyone's mind, was the knowledge that things were very wrong with farming. Farmers generally spent less and less time and money in sport and pleasure, and what pleasures we did indulge in had lost their savour. In this I refer to the older men. The young generation, from eighteen to twenty-five, knew nothing about the tragedy in the background, and seemingly cared less. I had a friend stopping with me in the summer of 1927, and took him one afternoon to a tennis tournament in a nearby village. Ostensibly it was in aid of a new church organ or similar object, but in reality it was an excuse for the youth of the district to have a good time.

'But where's this agricultural depression you've been telling me about?' asked my friend, as he gazed at the array of cars and at the expensively garbed youthful throng on the tennis-courts.

'It's there all right, in the background,' I said. 'These kids don't know anything about it, but their parents do.'

We sat and listened to the chatter. Said one young damsel to another: 'My dear, a frightful thing's happened. My new racket's got a string gone. Just my luck. Whatever shall I do?'

'Well, you've got another with you, so you can manage,' replied her companion.

'But, my dear, I shan't be able to hit a ball with it. It's last year's.'

My friend looked at me, and we got up and moved away. 'Why in the devil don't their parents stop it?' he burst out. 'Why, dash it, my racket's five years old.'

'Don't ask me,' I said. 'It's beyond me. Pride chiefly, I expect.'

And then for five minutes he trumpeted his opinion of the farming population, and in most of it I felt bound to admit that he had reason.

But the depression was there. Farmers knew it, the older labourers knew it, the banks knew it, and, above all, the merchant or middleman knew it. . . . And here I would like to put in a word in defence of the middleman. It has of late become the fashion to look on him as a parasite on the community. Farmers will, I think, agree with me that he is nothing of the kind. Personally, I have yet to find a middleman who is not performing a useful service for society, and all that he can earn thereby, in my opinion, he is entitled to keep. Moreover, for the past ten years right up till to-day, the agricultural merchant has carried, and is still carrying on his shoulders, a large proportion of British farming which, but for him, would collapse ignominiously. I have lived and farmed through those troublous times, and I am grateful enough to the merchant and middleman to say 'thank you' publicly for the help and courtesy which I have received from them.

By 1927, which, in addition to falling prices was a bad year for weather and crops, the rotten state of things in the agricultural world became more apparent. Here and there men who were regarded as wealthy by most folk, went bankrupt. Many others who escaped this were forced to give up farming because of financial difficulties. Farmers were endeavouring to get their sons into jobs unconnected with farming when they left school. A bank or Government appointment was looked at as a safe haven for life. It became preferable to let one's son do anything or even to do nothing, rather than to finance him in any farming venture. The landlords suddenly discovered that there was grave danger that the farms becoming vacant would have to be farmed by themselves. No one seemed to want them. So rents went down a little, but they were still above the pre-war rate.

I was struggling on, barely keeping my head above water, when early in 1927 a neighbouring farmer invited me to go with him to the north of our county in order to inspect a new milking invention. I refer to the open-air system of milk production, which was then in its infancy. Five of us made the journey in a car, and discovered this invention in what may honestly be described as 'Heath Robinson' condition. Quite frankly, we regarded it as a joke.

It was constructed of odd wheels and parts taken from derelict farm machinery. The milking cows were under-sized, and to our ideas not worth having in a dairy. But the thing worked, and the cows produced milk. We saw a man and a boy milk and feed seventy cows in two hours that afternoon, with one of these outfits. In actual fact we saw, I think, three outfits at work. And above all, we met a man who was satisfied that his farming was prospering.

Still, we looked at his cattle, and remembered our own wealthy beasts at home. We thought of having one's milking dependent on the spark from one magneto; we considered the difficulties and discomforts of milking in the open air during bad weather, and while we were very interested, and said so, we went in to tea with the inventor, feeling that this sort of thing could never become a general practice.

Now all that afternoon we had been literally surrounded with milk, but when our host said to his daughter that we were ready for tea, she informed him that it would be necessary to wait for a few minutes as the milk had not yet come down from the fields. This afforded us great amusement and one of the company informed the lady that had he known that they were short of milk he would have brought some down with him.

After tea we discussed costs and milk-yields with our host, and after thanking him for a most interesting afternoon, we set off for home. It is interesting to record that of the five of us who visited that farm on that occasion, four are farming under that system to-day, and the fifth, I think, will soon be doing so.

Now all that year I kept on worrying over this new idea, chiefly because my present methods were losing money, and something different had therefore to be done. I journeyed to the inventor's farm several times in company with my neighbour. In the June of that year the latter purchased an open-air outfit, and, as he farmed nearby, I was able to get many opportunities to study it.

He paid me a compliment one day by informing me that I wasn't quite such a b—— f—— as many young men of his acquaintance, in that I had realized for some

227

time that farmers were living in a fool's paradise, and also that I had made some effort to put things right.

'What about it?' he continued. 'There's more to this outdoor business than appears at first sight.'

'I'm inclined to agree,' I replied, 'but I've tried so many things during the last few years, and found them disastrous, that I'm getting afraid to trust my own judgment.'

'Well,' he said, 'for your information, I'm going in it for whole hog. I've been milking over two hundred cows by hand in buildings, and it's got to stop. My pocket can't stand it.'

Now while it was a comparatively easy matter for a wealthy man of his type to change his whole system of farming, for a young man already at his last financial gasp to do so was a difficult matter. But the returns from the 1927 harvest and milk sales left me with no alternative. It was try something else or get out as a failure. True, it would be a leap into comparative darkness. I wondered if it would be a case of out of the frying-pan into the fire. Still, the frying-pan had become untenable, and as this new system of dairying seemed to show a possible way out, I decided to try it.

CHAPTER XX

Now my farm of about seven hundred acres was in reality two farms, which had been farmed as one holding by my father for some twenty years prior to my tenancy. While it was possible for me to carry it on chiefly as an arable farm for a few more years before finally crashing, to go in for dairying solely on the whole of the farm was impossible for financial reasons. There was only one farmhouse on the two farms, so at Michaelmas 1927, I gave notice to quit the outlying farm in 1928.

The next thing was to plan for the new system. Firstly, I had to arrange the cropping on the farm which I was giving up, in proper proportions of corn, hay, and roots for the incoming tenant. Secondly, all the arable land on the home farm had to be put into corn and the grass seeds sown in this crop. I worked it that the major portion of this land was put into wheat, so that I should have a goodly store of straw for thatch in years to come.

Some description of this new system of farming is necessary here, I think, although to-day we have all got so accustomed to it in this district that we regard it as a matter of course. I will be as little technical in this description as possible.

The main point about the whole business was that one specialized in milk production, and nothing else, having no arable land at all. The cows were milked out of doors by mechanical means. A shepherd's house on wheels housed the power plant for the milking machine, and also an electric lighting set. The cowshed was also on wheels, and possessed no floor. It contained stalls for six cows, and was fitted with a mechanical milker. There was also sufficient movable chestnut fencing to make an enclosure or corral in which to assemble the herd at milking time.

The most obvious advantage of this new system was of course in the saving of labour for the actual milking. Instead of six milkers for sixty cows, only two were required. But there were many other advantages, which, while they were perhaps not so obvious, were even more important.

The cows were milked in the field in which they grazed, the outfit being moved on day by day to a fresh clean site. The cows were folded over the land almost like a flock of sheep, thus manuring the pastures. Instead of one having to cart the hay to the buildings to feed it, and then to haul the manure back to the fields, the hay could be stacked and fed in the field in which it was grown, while each and every cow became her own manure cart. This enabled one to farm, and, instead of impoverishing the land as by grain-growing, to enrich and improve the pastures year by year.

The cattle were definitely benefited in health, in cleanliness, and in temper. In health because of the sanatorium conditions of their open-air life all the year round : in cleanliness because they lay on clean grassland instead of in dirty, mucky yards or buildings, and in temper—an important point this—because they were never driven along roads, and hustled into buildings to be tied up. In this paragraph I have referred to buildings as dirty and mucky. What I mean is that buildings inhabited by cows will be dirty and mucky, if they are not cleaned out. Most dairy farmers do clean them out adequately, but this costs money to do, an expense which this new method did not entail.

Possibly that sentence about the cows not being tied up is a trifle obscure. The milking procedure was as follows. The herd was collected from the surrounding pasture into the chestnut paling enclosure. Six cows were then let into the six stalls of the portable shed or 'bail', and a chain was hooked round behind their rumps to prevent them from backing out. The concentrated feeding stuffs were carried in containers in the roof, and apportioned out to the cows into mangers by movable slides operated by the dairyman. The milk was drawn by mechanical milkers, then carried in a nickel pipe to the end of the shed, and finally deposited in the churn direct from the cows' udders.

When a cow was finished milking, the dairyman pulled a wire rope and a door opened in front of her nose, through which she walked into the pasture, to resume her life under natural conditions once more. The engine house and milking shed were lighted with electricity for the dark mornings of the winter, and there was a folding veranda outside the cow stalls to shelter the men in bad weather.

THE WANING OF THE GLORY

When my men came to know that I was giving up the larger portion of the farm and going in solely for outdoor dairying on the remainder, they disapproved very strongly. They had heard about this new system of dairying, and regarded it as idiotic. 'Mid be all right in summer time,' they said, 'but who's gwaine to milk outdoor in winter? Whoi, it bain't feasible.'

One man frankly said that the inventor should be strung up by the neck out-of-doors to see how he liked it, and also that he, the speaker, would be pleased and proud to haul on the rope on such an occasion.

I could sense that trouble was brewing, and began to find out which of the men were willing to have a shot at the new job. I had at that time a good foreman. He was hard-working, conscientious, and had always studied my interests, but his heart was in ploughing and corn-growing, where, incidentally, is mine also. At that time I was paying him as foreman of an arable farm, fifty shillings weekly plus his house and various other advantages. I offered him the same money to carry on at the new job the following year, and to my great astonishment he refused, saying that he would sooner starve. 'Farming's done, I reckon,' he said, 'if 'tis come to capers like that. I shall get a job somewhere else where I can drive a tractor.'

The only man who viewed the thing at all reasonably was my old dairyman, but he was seventy years old, and therefore out of the question as an active participant in this new method. Still, he was cheering. 'I've a seed a main few things in me time,' he said, 'and I bain't gwaine to say as 'ow nothin's no good till I've a seed un tried.'

As I should want but very few men under this new

system, I decided to purchase the outfit in the spring of 1928, while I still had the full staff of the two farms, and see which men showed any aptitude for it. I had at that time a dairyman with a family of sons milking my large dairy herd at the home farm, and as I should now want but two milkers, and he wanted work for about five, he would not entertain the new job, which was understandable.

There is a streak of obstinacy in me somewhere—I have yet to meet a farmer without it—and these repeated refusals made me the more determined to go on with the business. 'Blow the men,' I said to myself. 'I'll do it without them.' Accordingly, I gave the dairyman and his family notice to leave at Lady Day, March 25th, and took a pupil, finding him in board and lodging in return for his work.

This succeeded, admirably. The new outfit arrived in March, my pupil came a day or so before, and the dairyman and his family departed to a fresh job as arranged thus reducing my labour bill by about eight pounds weekly.

My pupil, Walter Bailey, was a townsman of about twenty, and he had never had any intimate dealings with cows in his life prior to coming to me. I imagine that he had only seen cattle grazing at a distance from the roadside. Thinking I might as well be hanged for a sheep as a lamb, I made him head dairyman for the new outfit. This was not so silly as it sounds. Granted, he knew nothing about cows, but he knew as much about the running of the new outfit as any of us, and that was precisely nothing. But he was eager to learn.

We began one afternoon to break in some forty-seven

cows to the new machine, which we had hauled into a forty-acre pasture. We should have been better advised to have placed it in a smaller field, but this we were not to find out until the following morning.

That afternoon was a picnic. All cows have minds of their own, and these decided that the new shed held no attractions for them. Some were driven into it, others pushed in, and a great number had to be literally carried into it. Strangely enough, they did not object to the mechanical method of milking, and stood quietly enough as a general rule while this business was proceeding. I imagine that they were too scared at the strange experience to object.

But we got precious little milk. This herd of cows had been regularly recorded, and the previous twenty-four hours they had given ninety-three gallons of milk: fifty-eight gallons in the morning, and thirty-five in the afternoon. That afternoon we got only seventeen gallons from them, as they refused to let their milk down, in the same way as they refuse to do so to a strange hand-milker. However, by dint of much coaxing, allied at times to brute force, to say nothing of its usual companion— 'b—— ignorance', we got them all through the performance, and went home to our teas.

Walter Bailey must have found the next morning a weird experience. For a town lad to rise at 4 a.m. and to wander out into a dark wet world inhabited by angry cows and still angrier human beings must have been rather terrifying. Of course, it was raining. During many years' experience of agriculture I have always found that whenever one wanted a particular type of weather on a particular day for a particular job, the Clerk of the

Weather invariably ordained exactly the opposite type. I suppose that this occurs just to point out that it is arrogant and presumptuous for any mere human being to arrange the night before to do anything next day dependent on the morrow's weather conditions.

We started at 4 a.m., as we needed to start milking at five, and we had anticipated a little difficulty in getting the herd into the enclosure. A little difficulty! Ye gods! The labours of Hercules must have been child's play compared with it.

There were about six of us, including a man sent to instruct us by the makers of the outfit. We found the herd and proceeded to drive them towards the enclosure. They went quietly enough until they got quite near to the outfit, and then they stampeded in all directions.

If there was one corner of that pasture to which those cows were not going, it was the one which was occupied by the outfit. 'Anywhere but there' was their motto. Why should respectable aged ladies be subjected to these new indignities? Where were their old comfortable buildings in which each one had her particular stall by right of long tenancy? And who were these fools who persisted in attempting to drive them away from their old home?

I can still visualize it all quite distinctly. Two bobbing lights moving to meet each other, as their carriers ran in a futile endeavour to stop the cattle. Dark shapes of cows on the skyline, streaming away between the lights, with their tails waving high in derision. The vapour of their breath in the lantern-light, when, after rushing straight towards us, they wheeled with hoarse snorts of rage and disgust at a few yards' range. The squelch of one's rubber boots, the tumbles at full length on the wet grass, sodden endeavours to relight one's lantern when this happened, the sound of bad language floating up from all sides through the wet dark, and over all the rain, the persistent steady rain of southern England, which not only damped us physically, but which sapped all our courage and endeavour.

And then, just when we had almost decided to wait for daylight, a final attempt succeeded, and we started the engine and began milking.

There was comparatively little difficulty in getting the cows into the shed when their turn came as compared with the previous afternoon's performance. The memory of an abundant feed of cake in this strange building was still fresh in their minds. Also they decided to give down their milk, for that morning we got seventy gallons, and went home well pleased to a substantial breakfast.

In four days the milk yield reached its pre-machine standard of ninety-three gallons, and in a week's time it had gone up to one hundred and one gallons. Walter took to the job as a duck to the water. I put a young dairy lad with him to strip the cows after the machine, and in a fortnight the two of them managed the whole job without assistance. In a very short while the chief difficulty was to keep the cows from forcing their way into the shed before a vacancy was ready, and a gate had to be rigged up between the bail and the enclosure to keep them out until their proper turn.

That year, as I remember it, was a fairly good season. Grain prices were down again, but the drop did not take place until about October. As I was giving up grain-growing, I threshed everything out of the harvest field in order to be able to sell my threshing machinery at my outgoing sale in October, and by doing so I made a fair price on my grain.

A farm sale always appears to me as a depressing sight. Whether its cause be death, sickness, failure, or even success, there is always the sense of a painful uprooting in the background. Ploughs, binders, and other implements, which often look majestic while in the performance of their proper duties, present a mournful appearance when exposed for sale in monstrous, gaunt, unsightly rows. Horses and other livestock are not simply business stock-in-trade, but intimate companions, with whom one has lived and toiled, with whom one has suffered defeat and achieved success. To see them all exposed for sale under the auctioneer's hammer, to see crowds of keen buyers hunting for their imperfections, and, above all, to admit publicly that one cannot carry on, hurts, and hurts badly.

But I had a good sale, although, as I was keeping my cows, I had chiefly only implements to offer. The utter failure of corn-growing with its attendant grassing-down policy had not yet occurred, and there was still a fair demand for ploughland implements.

However, in spite of this fact, I had a good sale, chiefly, I know, through the good offices of my friends, for which I thank them. There is, I know, a general impression abroad that farmers are self-seeking, grasping folk as a class. This idea has been fostered by many books during the last twenty years. To depict a farmer as a ruthless tyrant, oppressing his employees, and cheating his neighbours, seems to be a popular method. But my own experience with them, both in this country and in Canada, has been a very pleasant thing. I have always found my farming neighbours ever ready to give a helping hand, and I can say quite truthfully and frankly that without the help, advice, and sympathy of one or two of my farming friends in this country, I should have been down and out long since.

In looking back on my life I must confess that the bulk of the worth-while things in it are connected with friendship, which to me is all the more strange, when I think that I have lived my life chiefly struggling for material things, as do most of us, I fear.

Walter Bailey, who had run my milking outfit admirably that summer, came to me in July, and suggested that he was now competent to manage a farm of his own. After consultation with his father, I found a suitable small place for him, and he started on his own there that Michaelmas with seventy heifers, a milking outfit, and two men, after but six months' experience. He had learnt this new milking business, and he had seen one haymaking, but there was nothing else to it, as his farm was all grass.

It is interesting to record that he made a success of things after such a short training. On a capital of about twenty-three hundred pounds he paid five per cent interest and made a handsome profit during his first year's farming, and, for aught I know to the contrary, he is still doing well. His success was due to his own personality, to his capacity for work, and to his tenacity of purpose, but in a small way I take to myself some little credit for it. He benefited from the mistakes which I had made and paid for during the previous years.

CHAPTER XXI

Just prior to my sale I planned out my future farming campaign, especially as to which men and implements to take with me. I reviewed the position honestly, and found that it was worse than I had imagined. I thought of my criminally foolish extravagances of some years before, and came to the conclusion that there was no bigger fool in all the county. It was particularly galling to fail now that I felt convinced that I had at long last discovered a possible way out. But it was the back debts of my ploughland farming which were beating me.

Still, it was no use worrying about past mistakes, and the only way to combat the future was by hard work, and rigid economy. Golf had long become only a very occasional game on a Sunday afternoon, but even that must cease. Tennis and shooting also must be given up, not partially but altogether. It was no use messing about at economy, so I sold my golf clubs and arranged for a rabbit-trapper to catch my rabbits at so much per couple, instead of inviting merry parties of my friends to shoot them. Now, if I was not going to play at all, I should have to do something or I should get into mischief, so the best thing would be to tie myself down to a job of work which should take all my waking hours.

I should therefore need very few men. It was a beastly business saying good-bye to many old and trusted employees, but it had to be done. I had about a hundred cows, and should need two hands for them besides a general labourer for the water meadows, thatching, hedging, and other farm work. I arranged with a young man who had just been demobilized after service abroad to take on the milking outfit. He had no previous knowledge of dairying but was willing to learn and, after my experience with Walter Bailey, I felt that willingness to learn was the most important qualification required for this job. I kept on my 'drowner', who was willing and able to work on the milking outfit in emergencies. I should require no carters, as my remaining two horses were to live outdoors with the cows, their only work being hauling hay and cake for the dairy. I myself would be the other man on the milking outfit, and that would be that—a small but efficient staff.

There would be no work other than dairying until next year's haymaking, and that bridge could be crossed somehow when the time came. The foreman had gone to his new job, and the remaining men must take their chance of employment with the incoming tenant on the farm which I was quitting. The groom-gardener would be unnecessary. He was heartbroken, but there would be no place for a groom-gardener in this new scheme of things. My farming from now on had to be cold-blooded business and nothing else; there would be no horses to groom, no car to clean, and the garden must take its chance.

THE WANING OF THE GLORY

But I found a job for my old dairyman. This was ninety-nine per cent common sense and perhaps one per cent sentiment. There were a small number of the older cows who did not take kindly to the mechanical milking after many years of the hand method, and it would pay to have the old man to look after them. Besides his knowledge of cattle would be valuable. Anyway, he was over seventy, and if he stopped work altogether, it would break his heart and finish him up in a very short time. And, if any further reasons are required as to why I employed him, well, just because. Anyway, he came over to the home farm with me, and proved a veritable tower of strength at all times, especially in times of adversity.

And in this fashion we settled down to that winter's work. It was almost a return to my Canadian life. I rose at four-thirty and milked, returning home to breakfast about seven-thirty. From breakfast until dinner time Dick Williams, my new ex-soldier dairyman, and I, hauled hay out to the cows, moved on the outfit to a fresh site, and did other necessary jobs in connection with it. After dinner we milked again, fed the cows their evening allowance of hay, and returned to the buildings. The old dairyman milked his few cows, washed the churns, and cooled the milk ready for dispatch by rail to London. Dick's last job was to take the milk to the station, whilst I would go over into the meadows to see how Bill Turner, the drowner, was getting on with the irrigation work.

And that was our life that winter; a daily round, a dreary monotonous furrow if you like, from which there was no escape either on weekdays or Sundays. But the fact that it could not be dodged was its strong point.

241

THE WANING OF THE GLORY

Instead of our work depending on the weather conditions, the weather was ignored. The farm had become a factory. In good weather it was pleasant to do our job, in bad weather it was unpleasant, sometimes very much so, but always our job was done. Every time I paid Dick his weekly wage of forty-five shillings, he had, with my help, milked my cows, fourteen times, whatever the weather conditions might have been. I was paying only for services rendered.

The old dairyman had done much the same with his little herd, and as the drowner was doing the meadows at a piecework price, I was only paying out for profitable work. Therein lies the strength of grass farming of any kind. Contrast it with the arable method. Carters and horses idle perhaps for a month or more owing to frost, and similar hindrances and wastages caused by weather conditions throughout the year.

Of course, it was not all honey. We had many setbacks and vicissitudes of fortune to overcome. An oil engine does not always start, or at least it does not do so until you have found out the reason why by bitter experience. It is dark until 7 a.m. on winter mornings, and electric lighting sets are not infallible. When one wrestles personally with a faulty connection by the light of a lantern in a hurricane of wind and rain, one sometimes wonders if it is worth continuing the struggle.

But the job always got done somehow. It is wonderful what one can accomplish when there is no possible hope of getting any assistance. I think that everything that can happen to an outdoor milking outfit happened to mine that winter. We smashed the under-carriage of the engine house whilst hauling it through an awkward gate-way. We arrived one morning to find that the wind had blown the milking-shed upside down, owing to our neglecting to shut down the veranda. We had it frozen up many times, so that although the engine ran merrily enough, the milkers refused to function. We barked our fingers, cows trod on our toes—rubber boots are small protection against this painful happening—we swam at times in a sea of mud, and froze in an east wind on other occasions. But each and every day the job was done, and done at small expense.

Generally speaking, the cows gave but little trouble. They were eager to come in at each milking time, for in the bail was their cake. They became extraordinarily docile, and for the most part completely ignored the milking machine, giving down their milk immediately it was attached. The bull ran loose with them, and he also became quite docile. I expect that we lived so much with these animals that we took them as a matter of course. Sometimes when one left the glare of the light and went outside the shed to open the gate for another cow to come in, Billy, the bull, would loom majestically out of the dark, unconcernedly push by you, and take his place in the bail. I am afraid that we did not let him occupy a stall for long enough to eat a decent breakfast, as we had some seventy cows to milk, and could not waste time with him. So the door in front of him would be quickly pulled up, and William would be prodded forth into the pasture once again to make way for one of his lady friends. He used to grumble to himself at this treatment, and would presently come rumbling round to the end of the shed and try to force his way in once more. When one of us went out to see if the churn was full and wanted changing, William would be sure to be hanging about there in the way, but we had no time to take any notice of him. A slap on his nose with the bare hand was usually all the attention he got.

It is rather curious to think that the milking and feeding of some seventy cows depended on only two men. At 4.30 a.m. on a winter morning the great majority of our population are in their beds. Very certainly the Wiltshire downs at this time of day are not thickly populated. As I walked to work each morning I seemed alone in an uninhabited country. Then far ahead a light would suddenly spring into being; evidently Dick had arrived all right and had just switched on. In a few moments I would blunder into a cow. Having prodded her into a slow and disgruntled walk towards the light, I would scout around and perform the same office for her companions. Very reluctantly they would struggle to their feet, and slowly amble towards the light. Ah! What was that? The 'put, put' of the oil engine commenced to chatter cheerfully in the distance. At this the cows abandoned their reluctant, sluggardly pace for a brisk walk. They had heard their breakfast bell, and now had a purpose in view.

I had been warned that it was inadvisable to put old hand-milked cows on to a milking machine, and that January I was to find out the penalty for so doing. We had a slight outbreak of mastitis or inflammation of the udder. This often occurs in hand-milked indoor herds, but it is more easy under those conditions to deal with it. Isolation of the affected cattle is obviously the best method. With a milking machine under our conditions, in spite of every precaution, it spread quickly.

THE WANING OF THE GLORY

The only way out was to take away the old herd and begin again with first-calf heifers, which I should have done in the first place. First-calvers, having never been hand-milked, take to a milking machine more easily than do old cows, and, considering that method to be the natural order of things, give no trouble of any sort. I engaged another milker, and with him, the old dairy-man, and the drowner, Dick and I started hand-milking once again, and found it a dreary, dirty business. I sold some of the cows as barreners as soon as possible to reduce the numbers of the hand-milked herd to reasonable proportions, and also to get some money with which to buy some heifers. By April we had them reduced so that the dairyman, the drowner and the new man could manage them.

Meantime we had thoroughly cleaned and sterilized the outdoor plant, and in May started it going once more with fifty Irish heifers. These cattle were purchased from a dealer in Ireland as springing to calve with their first calf. They arrived at our local station looking very sorry for themselves, but in a few days they improved out of all knowledge.

The arrival of one lot of these cattle gave great heart-searchings to our local policeman. They were all imported under licence, on which was stated the number and class of cattle together with the different numbers of their ear tags. This licence had to be given up on their arrival to the policeman, who then had to inspect the cattle and check the number. On this occasion the licence said that sixteen cattle were to arrive, and we unloaded seventeen from the trucks, as one calf had been born on the railway somewhere between Fishguard

245

and our local station. Accordingly the policeman was worried. Sixteen cattle were licensed to arrive at our station, and seventeen had been detrained. His opinion was that a breach of the regulations had taken place, so the law must take its course.

Dick and Bill, who had gone up to the station to get the cattle, argued that there were sixteen Irishmen and one Englishman, regardless of sex. 'Nice little veller,' said Bill, fondling the calf which he had placed in the milk float. 'He wor born under the British flag, and I be takin' 'is part as a trueborn Englishman. I 'low if thee's been born like 'ee wor, thee ussent never a wore no blue trousies.' Subsequent inquiry to high officials proved that Bill had the right of the matter.

I hope that the inventor of these milking outfits will not be annoyed at my telling of that outbreak of mastitis and its consequent expensive reorganization, but I have tried to tell the truth of the matter, and can honestly say that I have had no further trouble since that time.

By this time haymaking was imminent. I had one hundred and eighty acres of hay to cut and make, chiefly new leys, which had been sown in my last year's corn. Three of my staff were busy with the old herd, and Dick and I had the new heifer herd to manage. It looked as if our hands were full enough without any haymaking to do. But it had to be done, so I put the drowner to help Dick, and said firmly that the other two must manage the old cows somehow. There are times when one must say firmly that things must happen, and not wait for any discussion as to their possibilities.

I had kept my tractor when I gave up the other farm, and now purchased a power grass-mower with a seven-feet wide cut to attach to it, making a one-man mowing outfit. I drove this from 5 a.m. until about 8.30, when Dick returned from his breakfast, and took on while I had mine. He drove until 12.30. I had an early dinner, and kept at it until he returned after tea, from which time he drove until 8 p.m., my last lap being from that time until ten o'clock.

When the first day's cutting was fit to pick up, we kept the mower going from 5 a.m. until the dew was off the hay, then attached a hay-sweep to the engine, and went on carrying until eight o'clock in the evening.

This daily cutting is important. Nothing spoils hay more than sunshine. If you get a large acreage down, and rainy weather comes, everyone knows that it will be spoilt. What is not so generally known is that hot sunny weather will bleach a lot of it to the feeding value of straw before you can get it picked up.

For ricking staff I got four men from the local labour exchange for three weeks. I think it best to give but few details of their performances in the hayfield. One was an old farm hand, two did their best, and one was hopeless. Still, I was only left alone with them during the after-noon milking, and I was very lucky with the weather. That summer of 1929 was an exceptionally dry one, and we made a moderate crop of hay in first-class condition very cheaply. Incidentally, I am employing the same methods during this haymaking of 1931, but Providence has not been so kind. Still, we are almost finished, al-though it is pelting with rain outside as I write.

Haymaking finished, we settled down once again to our factory routine, and by the Michaelmas of that year the number of old cows had been reduced to a point where the old dairyman could handle them, with some assistance from the drowner in the morning milking, so I was able to dispense with the extra hand-milker.

At this time, I began to realize that this grass farming was an uninteresting business as compared with the older ploughland system. The romance of farming was gone. There was no seed time, and apart from haymaking, no harvest. There was no plotting and planning ahead for future years. The few men I employed were engaged in a factory-like occupation day after day. The landscape of my farm presented a dull, green sameness throughout the year. The glorious patchwork of different kinds of grain crops, alternating with green fields or roots, and here and there a brown fallow, was now an expanse of prairie. Apart from the old dairyman pottering about the buildings, the steading was deserted. The stables were empty, and the gardens and grounds of the house were neglected and unkempt. Still, regarding this change as I did, I was glad that I had done it, for corn was going down and down.

That there was resentment in the farming world at this new method of dairying cannot be denied. There is no disguising that its advent and increasing use caused great bitterness. Its advocates, though then few and far between, were enthusiastic. Its enemies were many and vindictive. This was understandable. Here were men on cheap hill land saying that they could produce milk at several pence less per gallon than could the older established dairy farmers on expensive valley land. The latter were in much the same position as were the stage-coach owners at the advent of the railway train. The greater production of milk, due to corn-growing farmers

248

abandoning their old methods and engaging in dairying, caused the price of milk to fall a shade, and this was definitely laid at the door of the outdoor system, which rendered it easier for the arable man to change his mode of farming. This bitter feeling was the more acute in our district, as, for obvious reasons of climate, this outdoor method can only be generally practicable in southern districts.

As no record of farming can be completed without some reference to the National Farmers' Union, I would here say that that much-maligned body fulfils a useful purpose in our national life. Like most institutions of this character, its chief difficulties are its own members. Farmers are naturally individualists and do not take kindly to control or to co-operation. Their interests are diverse, not only as between county and county, but even between neighbouring farms. To present a united front to any Government on a national question is therefore a difficult job, which the Council of the Union accomplish far better than is usually supposed.

Looking back on my own personal experiences while serving on a County Branch of the Union, I find them very curious. Whenever I voted with the majority on the popular side of any question, after-events have proved me to have been in the wrong, whilst on the rare occasions when I have voted with a very unpopular minority, after-events have proved my action to have been the correct one. Possibly most public life of this nature works out in about the same fashion.

But the National Farmers' Union is a responsible body with which any Government can discuss agricultural problems, and, as such, justifies its existence, whatever its many critics may say to the contrary. Naturally there are all types of men who take a leading part in it. Many of these may be justly described as seekers after personal glory and power, but thank heaven, there is a leaven of broadminded common sense folk running right through the N.F.U., from Northumberland to Cornwall. These are seldom popular as they often point out unpalatable truths to their fellow members, but although they are usually disparaged by many farmers during times of prosperity, it is to these somewhat unpopular folk that the main body of farmers turn, when they are in a bad hole.

Of my small staff at this time I can speak only with grateful appreciation. Dick Williams never grumbled, however unpleasant the weather conditions might be for working the outdoor milker; Bill Turner, the drowner, was a willing jack-of-all-trades, and the old dairyman, George Strong, was a Rock of Ages, both in appearance and reality.

It will be a bad day for British Agriculture when the older types of farm labourers are no more. They are lessening in number rapidly in these days, but usually there are still one or two to be found on most farms. They draw their legal wages for day work, but infinitely prefer to work by the piece. Employees they may be, but I prefer to call them quite sincerely, 'Guides, philosophers, and friends'.

We had a wet time in the autumn of 1929, and the gravel drive leading up to the house became green with weeds. My wife, who by this time had become resigned to an untidy garden, said firmly that the drive must be weeded. I was not surprised. She had cheerfully put up with a good deal during the past year, but this was the last straw. Struggle as she might to keep the garden passably tidy in her spare time, to weed a long gravel drive was out of the question. And she had spoken firmly you note. Any married man will realize the seriousness of the situation. Something had to be done.

Of course, I might have done a little weeding personally in between milking, but I do not like weeding as an occupation nor as a hobby, so I cast about in my mind for a way out. The only possible available man was Bill Turner, but he was the drowner, a craftsman of high degree. I considered that gravel-weeding was beneath my dignity, and also that it was beneath his. He would do it if I told him to do it, but somehow I did not quite like to take this course.

However, the difficulty was solved without my aid. I wandered over into the meadows next morning, and communed with the drowner—deep calling to deep. We talked of stops, of drawings, of hatches, of tail water, and other important technical matters. Incidentally I mentioned that the meadow work was well forward for the time of the year.

'Ay!' he said. 'It be. We'm got 'em underhanded thease season. But I do want to lave 'em fer a few days. Thic gravel on the drive be got main weedy. Do look reel bad. I doan like to see it.'

A day or two afterwards, when I came home from the afternoon milking, the gravel was being weeded. I went out to cheer on the weeder. I may be a poor weeder, but I am a good cheerer-on—a more valuable quality than many folk imagine. We discoursed on the weather and other serious matters for a while, and then he pointed out two paths, each about four inches wide, which curled away round the house in opposite directions.

'Do 'ee know who do make they?' he asked. I examined them carefully, but could not find no satisfactory solution as to their origin.

'That be Trinket, that be. She do listen fur 'ee like a cat fer a mouse. When you be in dinin'-room, she do squat outside the front door. If she do hear your step goin' to back or side door, she do trot round thic way. When you do go upstairs to shave er zummat in the bathroom, she do take t'other road, and sit under the window till you be done. I bin a watchin' she to-day. Faithful old bitch, she be.'

I agreed with his last remark, and took particular notice afterwards to find that he was right in every detail.

Poor old Trinket had been having a dull time since I had given up shooting, and I had been taking her selfless devotion as a matter of course. Who am I to be loved so well? A man may not be a hero to his valet, and few of us, if any, are heroes to our wives. They know us in all our moods, when the limelight of public opinion is not shining on us with its merciless glare. I hope that I may be considered somewhere near average as a man, but nobody thinks that I am a fine fellow in the same way as does Trinket, my golden retriever.

And what do I give her in return for this devotion? Precious little. An occasional pat on the head, a few words of greeting when we meet, permission to come indoors and sit by the fire with me when the powers that be are not about, and some fairly daft one-sided conversations on these rare occasions, accompanied by much ear-pulling and tickling her under her collar. It is true that we occasionally go out and slay a rabbit together, but that is the serious business of her life, and not to be confounded with these gentler relaxations.

In due course the gravel was weeded, and about once a year we continue to make an effort to keep it respectable. But usually the drive presents a green appearance to the caller instead of its spotless golden brown of years gone by. Well, I do not greatly care. Trinket is now seven years old, and, in the natural order of things, is getting a trifle slow, and more than a trifle deaf. Though I do not like to see weeds on the gravel, they cause me no real grief, but I shall be very sorry indeed when the day comes that I do not see those little paths.

CHAPTER XXII

During that winter of 1929 I found the situation to be decidedly better than it was a year before. This new method of dairying appeared to have stopped the leak in my farming ship, but, to continue the metaphor, there was still too much water in the boat. Stopping the leak was not enough. The dead water must be pumped out somehow. To find an efficient pump for this seemed impossible.

Now when my old dairyman was managing the small dairy on the farm which I had given up, he had done a little milk-retailing in the small village in which this farm was situated. This retailing had been carried on even in my father's day. It had been in no sense ordinary retail trading. There were only about a dozen cottages and one large house in the village which last was occupied by a retired Admiral. Most of the folk came to the dairy for their milk, and the old man toddled round to the others, including the Admiral's. The accounts were kept in chalk on a child's slate, which was hung on a nail in the dairy, and the takings were paid over by the dairyman on each weekly pay-day.

This had always proved useful, as the coppers thus collected were used to pay the odd money in the men's wages caused by the insurance deductions. The dairyman used to read the amounts from the slate to me on pay-days, and after a while I noticed that the Admiral's account never varied. I commented on this regularity one day, and the dairyman said that he had told the Admiral that he must have a regular quantity or none at all.

I knew that the dairyman was an autocratic old bird in his way, but the Admiral's reputation was certainly not renowned for meekness, so I inquired particulars.

'Oh, I jist told un,' the old man replied. 'I telled un as 'ow I werden gwaine to 'ave no 'oppin about. A quart one minute and a pint and a half the next. I telled un he had to 'ave a quart mornin' and a quart night, as I wadn no clurk.'

'But supposing they don't want so much sometimes?' I asked.

'Oh, 'ee got plenty o' money to buy two quart o' milk, and 'ee all day to drink it in. Any'ow, I telled un as I wadn gwaine to 'ave no argyfyin' about it, and 'ee do 'ave it.'

That is, I think, an example of the most ideal form of milk-retailing. United Dairies, with all their modern methods, will never attain such a high standard.

THE WANING OF THE GLORY

But I could remember that those few gallons of milk, thus sold, came to quite a few shillings each week, being sold at a much higher price than the bulk of the milk which went to London. Why should not I try to sell my milk retail?

I played with this notion in my mind all the winter. The only possible market was in the country town some five miles away. Well, that would be only a ten minutes' journey with a motor van. The next point was whether I should sell enough milk to make the venture worth while. I would look an awful fool to have a go at this business, and afterwards to be forced to retire defeated.

But there was another seemingly insuperable objection. I should have to do this work myself, at any rate in the beginning, and I discovered that I was a snob, an awful snob. And yet, God knows, I had nothing to be snobbish about, but rather plenty of things of which to be ashamed. Still, to be a milkman, at the beck and call of any one who might purchase a pint of milk from me! What would people think? What would my friends think? Fancy donning a white coat and possibly creeping round to the back doors of houses, where I had been accustomed to enter as a guest by the front entrance. Fancy going round soliciting custom. No! I could not do it.

These and similar thoughts ran through my mind. I suppose that everyone is a snob in some way or other. That I was one over this matter, I confess. I am ashamed of it now, and at the back of my mind I was ashamed of it then, but at the same time it was a very real problem. That I knew, deep down, it was wrong and despicable did not make it any the easier, but I think, the more difficult.

And I knew that I had a good article to offer to the public. Milk produced by this open-air system is cleaner than a large proportion of milk which is milked by hand in buildings. I had proved this by bacteriological examinations, and by practical tests of its keeping qualities in my own larder. Moreover, the open-air idea as a cure for tuberculosis was a popular one for human beings, and cows which were never shut up together in hot buildings, were surely less likely to contract this disease. That point should weigh with the mothers of babies.

The more I thought about it the more convinced I became that there was a chance of working up a retail business. Well, was I in a position to ignore any possible chance of improving the returns from my farm? No! In justice to my creditors, and if I was to retain any semblance of self-respect, I must make an attempt. Who knows, there might be some fun as well as some money in it? Snobbery must take a back seat.

I decided not to start until after haymaking, so as to have that worry off my mind for at least another twelve months, and when it was over, we commenced operations. I inserted in the local press an advertisement extolling the fine qualities of my goods and announcing that I should commence retailing them in a fortnight's time. The next step was to send a circular in the same strain to all the people in the town, with whom I had any acquaintance either from social or business reasons.

Once again my friends turned up trumps. My scriptural knowledge is a trifle hazy, but I think that it was the 'unjust steward' in the parable, who made friends of the 'mammon of unrighteousness' during his palmy days. Very certainly, I had been an 'unjust steward' some years before, but I hesitate to describe the friends I made during that time as of the 'mammon of unrighteousness'. But they were good friends which I had made in the golfing, tennis, and social worlds, and while most of them were mildly amused at my new venture, they supported me right nobly.

In the beginning my farming friends teased me unmercifully. On the market day after my advertisements appeared, I was greeted with 'What cheer, Milky?' and similar remarks. One wag informed me that the only way to get business was for me to make love to all the cooks in the town, and he painted a lurid picture of my procedure in this direction to the great entertainment of his hearers. Still, this teasing was all very goodhumoured, much the same as I should have indulged in had any one else started a venture of this kind. And, here and there, amongst the older farmers came comforting, approving, and helpful remarks. Some of these had relations living in the town, and advised me to call on them to try and get some business.

I confess that I funked this calling on people in search of custom. For one thing it meant dealing with the ladies, for a mere husband has no place in the domestic purchase of his food, and I soon found that any direct reference to him, or any suggestion from him to his better half, spelt disaster to my cause.

Still, you do not get business unless you go after it, so this canvassing had to be done. I well remember my first morning at it, and especially my first call. I sallied forth with a list of likely buyers, and in great trepidation rang the bell of the first house on my list. When its jangling had died away, I was seized with panic. What on earth should I say to the lady? All the nice opening gambits, which I had so carefully rehearsed in my mind the evening before, seemed suddenly to be silly and feeble. There was a feeling of apprehension in my tummy. My hands were moist as they were many years before when I had knocked at my schoolmaster's study in certain anticipation of a licking. I hoped and prayed that no one would answer the bell, so that I could bolt for home.

But in due course I was admitted and nervously stated my case to the lady of the house. She listened to me courteously, asked one or two pertinent questions as to service and times of delivery, and then said: 'So you are going to try to get some money out of the consumer yourself, instead of whining to the government to get it for you.'

That, mark you, from a townswoman! It opened my eyes as to the true opinion held by townsfolk of the agricultural situation. In this case it was my good fortune that this lady was not too well satisfied with her existing milk supply, and she promised me her business, expressing the hope that we should soon play golf together once more.

Of course, it did not all happen like that, but everywhere I found people most courteous and friendly. Where they were thoroughly satisfied with their present milkman they quite rightly refused to change, and wished me luck. I discovered that I could never tell from which source business was most likely to come. Where I considered my chances rosy, nothing happened, and often where I thought it hardly worth while to call, good business dropped into my lap.

I engaged a young man to help me with this business, purchased a motor van, and on the appointed day we set out for the unknown, in my case in a Christopher Columbus frame of mind. The old dairyman helped us load up, and as we were moving off he said: 'God send 'ee luck.'

That day we sold five gallons of milk. In one week we had got it up to twelve gallons. All this trade was chiefly with my friends. Then we settled down to quiet plodding, and slowly but surely the business grew. We had a lot to learn, we made many mistakes, but never a week went by without a new customer's name appearing on our books. In three months we had got the daily sale up to twenty-five gallons. This may not seem a very rapid rise to some people, but when you remember that on the average it means six customers to the gallon, it was not bad progress.

I was informed that I had missed my vocation, and that I should have been a commercial traveller. This may be true, for I talked about open air milk so much at that time that I could do it almost automatically.

But I was extraordinarily lucky with that retail venture. Everybody I came into contact with seemed anxious to help me. In looking back on my own life I can remember few occasions when I went out of my way to help other folk, but very certainly people have spared themselves no trouble to help me. One does little in this world unaided, I find.

Not only did my town friends help me with their custom, but my farming friends aided me whenever it lay in their power. One man's advice I shall always treasure. When he found out that I was going in for bottled milk only, he warned me not to buy any bottling machinery until the size of the business warranted it. This he put at a daily sale of fifty gallons. 'Keep your money in your pocket,' he said. 'You can bottle milk by hand. It means work, but that won't hurt you. Work never hurt anybody. Get up a bit earlier.'

So at the beginning we worked. We bottled by hand, we washed the bottles by hand, and we got up early in the morning that summer to supply our customers with the morning's milk, cooled, bottled, and delivered by 7 a.m. This delivery of fresh milk only two hours from the cow undoubtedly brought us business.

There is no need for me to describe our efforts during the next year in detail. Running one retail business is very like running another. There were moments of triumph and satisfaction, and times of despair and fear, but week by week the list of names in our ledger grew longer. There was also a certain amount of fun in the business.

The joy of most forms of sport is obtained from the pursuit and capture of something or other. I have chased rabbits, fished for trout, participated in fox and wolf hunting, and stalked moose and elk. Still, I must confess that in the pursuit of the wily milk-buyer to his lair I obtained even more pleasure than from these other pursuits. In addition to the joy of the chase and the satisfaction in the ultimate capture, this form of hunting paid. As I have said before, I have a mercenary mind.

At Michaelmas 1930, when we had been retailing for three months, the old dairyman came to me, and suggested that while he did not wish to stop work

altogether, he wanted to ease off. I was not surprised;
we had all been going it during the last two years. By
this time all the hand-milked cows had gone, and I was
milking no cows other than the outdoor herd, which
had greatly increased in size. From the sale of the last
lot of old cows I was proposing still further to add to
it. I suggested to the old man that he might stay on in
his cottage and help us at odd times and when any one
was ill. But he wanted some regular work, and proposed
that I should still keep one or two cows at the buildings
for him to look after; saying that I needed a certain
amount of cream for the business, which he could
separate from their milk, and that he could help a little
with the washing-up of the dairy utensils.

'You gie I twenty-five shillings a week and me house,'
he said, 'and I'll do what I can. I don't want to gie up
altogether. I still wants to 'ave a interest like.'

We fixed it up as he suggested, and it worked admirably, suiting both parties. But in a few months
Nemesis descended on us in the shape of a Farm Wages
Board Inspector. All was well with the other men's wages
but in the dairyman's case I was breaking the law. Even
though he was over seventy an official permit must be produced to pay him at less than the legal rate for a man over
twenty-one years. The Inspector asked the exact number
of hours the dairyman worked in each week. 'Goodness
only knows,' I said. 'I don't. You had best ask him.'

'Can I see him without you being present?' he asked.

'Certainly. He's over in the dairy now,' I replied.

The Inspector returned from the interview with a
broad smile on his face. 'Couldn't get much out of him.
He said that he came when he liked and went when he

liked. It's obviously all right but you'll have to get a certificate of exemption from paying the legal rate on the grounds of age and infirmity. We'll put the hours down at forty-eight.' The old man was forced to submit to the indignity of a medical examination to ascertain if he was unfit for a full day's work, and when this was done an official inquiry took place in the local market town.

I drove the dairyman into town on the appointed day. As we were walking through the streets he seemed to be very old and feeble. He could still handle refractory cows, but modern traffic bewildered him. The inquiry was held in the Town Hall. There were present a chairman, a secretary, two or three other members of the Board, and the official representative of the Agricultural Workers' Union.

When the facts of the case had been read out by the secretary, the Chairman looked over his spectacles and said: 'George Strong?' Whereupon the old man rose to his feet and said: 'Zur?'

And there he stood, a rugged old peasant like a gnarled and twisted oaken monarch. That he was a better man than any of us in the room was patent to us all. The Chairman asked him if he had anything to say about the business in hand.

'I bain't gwaine to be interfered with. I does jist what I likes. I allus have all me life. I do know what I be worth bettern any o' 'ee yer, and I do know what thease job be worth, and I bain't gwaine to be interfered with.'

He sat down, and there was a beautiful hush for a moment or two. Presently the Chairman fumbled with his papers, and asked the Labourers' Representative if he had any objections to the certificate being granted.

'None at all,' was the reply, 'I think that this is a case in which there is no objection on any grounds.'

In due course the machinery of the Agricultural Wages Board produced the necessary permit, and we returned home.

I do not tell this with any idea of poking fun at the Wages Board and its regulations, but only to point out that there are two sides to every question. I well know that the passing of these regulations is entirely due to the tyranny and injustice of some few farmers in days gone by. But all my farming experience has been that the agricultural labourer has always known his worth, and been well able to obtain it from the farmer. Also the relation between employer and employed has been a friendly intimate thing, in which each party has had a sound respect for the other.

I came across this sort of thing only the other day as I was paying Bill Turner. 'Twelve and a half hours overtime, Bill?' I inquired. 'Thee and I an't surely come to 'alf hours, zur? I an't never bin paid for a 'alf hour eet, and I bain't gwaine to start now. Thee pay I fer twelve hours. I shall cop t'other somewhen.' I broke the law, and so paid him. Whether he has 'copped' his half hour from me or not yet I do not know, but I am sure that neither of us is losing any sleep over it.

These new regulations of hours and wages press hardly on agriculture, owing to the fact that they are factory regulations. Farms are not factories with roofs over the top. Farmers have always had to combat the weather in their business, and although some seasons are bad ones from this cause, a retrospect of any ten years of farming invariably shows that on the whole

THE WANING OF THE GLORY

Providence has dealt fairly with the farmer in the matter of weather. But when in addition to his struggles against climatic conditions, a farmer is handicapped by factory-like regulations, he is fighting an uphill battle.

Granted, the old system had its faults, but it worked, and produced farmers and labourers such as I have described in this book. The present system of bureaucratic interference not only does not work, but its products in the shape of farmers and labourers are not to be compared with their predecessors.

I am well aware that I am very small beer compared with my father and others of his type, and with very few exceptions the modern young labourer is of little value as compared with his father, either as a workman, a citizen, or as a man.

During the account of our early retailing struggles I seem to have made no mention of the outdoor milking plant. There is nothing to tell. That is, in itself, high praise. It functioned twice daily with the regularity of dawn and sunset. It gave us no trouble, and by this time the effect of its travels over the pastures, especially on the downs, was apparent. During the summer of 1930 I cut a heavy crop of hay on a piece of down land. It was a remarkable tribute to the effect of the treading of the cattle and the cake fed to them during the previous two winters. It consisted of a heavy cut of indigenous grasses with a dense mat of wild white clover at the bottom. In this case I undoubtedly reaped what I did not sow. I believe that something awful happened to the man in the scriptures who did this, so I fear the worst.

It is just a year now since I began retailing, and I see on looking at my ledger that we sold fifty-eight gallons daily during last week. I am a mercenary individual, and I am somewhat disappointed as I had hoped to write sixty gallons at the end of a year's trading. But still this last year has been a record of steady progress, while the previous seven or eight years were a dismal record of slipping backwards, so in reality I am well content.

The old system of farming as described in the early part of this book was one which placed the farmer in a yearly rut from which there was no escape. One did the same things, year after year, according to the season. Mistakes of one year were noted and avoided if possible in the next. You were kept so busy and interested that you had no time to think of much else, becoming narrow-minded possibly, but by experience a better performer each succeeding year.

So it has occurred to me that one farms better in a settled rut of some kind or other. The war unsettled a good many one-time stable things, and in farming it is the change over from one rut to another which has been so painful and expensive. The chief difficulty has been to find another rut or system that will pay, now that corn-growing has become a bad joke. It matters not whether one finally settles on grass, sheep, dairying, pigs, poultry, sugar beet, or some combination of these so long as one eventually decided on a plan of campaign and sticks to it through several seasons, always remembering so to staff one's farm that one cannot get away from it oneself.

In my own case I have chosen milk-production on the open-air system, allied to the retailing of the product. I cannot recommend it as a pleasing thing. It might truthfully be described as a miracle form of existence. For my manifold sins I am now the slave of empty bottles. What a life! And not empty bottles with a glorious past behind them, but empty milk bottles. Can one imagine a more dismal fate? I am the servant of milk bottles all day long. I fill them, deliver them, book them, wash them, sterilize them, and sometimes, it is true, smash them.

In addition to my own milk I sell a certain amount of Grade A Tuberculin Tested milk, which I fetch daily from a neighbouring farm. Each night, in company with dozens of empty bottles, I take the road. They are confined to wire crates so that they shall be unable to attack me, their slave. But they clatter and babble incessantly. The lights of the car seem to illumine a wet misty tube along which I drive, surrounded by shrieking bottles. I am alone in the world in a moving Tower of Babel, driving through a damp and dirty Channel Tunnel. Where are the County Councillors, and where the Roads and Bridges Committee? At every pot-hole and bump the incessant scream is changed to the clash of cymbals. I cannot hear myself think. I just drive on and on through this tube of never-ending cacophony, pursued by a pack of brittle demons in full cry. It is an unsubsidized Grand Opera and I am alone in the auditorium, alone with bottles, bottles rampant and clamouring.

On the return journey their clamour is muted. They are full, replete, gorged, satiated. This repletion subdues their screams to a dull gurgling and clinking. How like the human race! But methinks I prefer them in their empty resonant condition. Now they are filled not only with milk, but with an overweening conceit. They do not deign to converse with a poor mortal such as I, but they whisper amongst themselves incessantly. They now wear a head-dress which bears the approval of the Ministry of Health. Who would not be conceited wearing such a crown? It makes me feel most humble, for the only trappings bearing the Government's stamp which I am ever likely to wear will, most certainly, be adorned with broad arrows. But I detest conceit, so on my return I plunge these bottles into cold storage. That takes them down a peg or two.

However, I am still a slave of bottles, for my next job is to add up the number of bottles full and bottles empty, which have enslaved me during the day. And when the final task is done I have my supper, and then, and not till then, I become the master of one bottle, which is not a milk bottle. Dexterously I draw the cork. Carefully and slowly I pour the contents into my glass. Reverently I hold it to the light to gauge its colour and condition. Then with a grave but genial nod, as if to say 'All's right with the world,' I drink a silent toast to milkmen, great and small.

It is early days yet to say whether this daily rut of mine is the right one or not, but I think that I might almost say, after supper anyway, that I seem to see a faint sign of daylight coming.

The old 'Farmer's Glory' is gone, never to return, but is this faint flicker that I see after supper a prelude to some slight return of 'Farmer's Glory' which shall burn surely and steadily, if not with its old brilliance, for many years to come? Who knows? Certainly not I, but I feel justified in hoping, and that surely is something in these dark and troublous times.

July 1931.

EPILOGUE

EPILOGUE

It has been suggested to me that having attempted in the foregoing book to paint a true picture of farming life in southern England during the past twenty-five years, I should in this chapter try to sum up the agricultural position as it exists to-day. This is a dangerous ground. In a story one can let one's fancy have a certain amount of rein, embroidery is permissible, and erasure is tolerated, provided that the broad outline remains correct.

But for a tenant farmer to write his own personal views concerning the present deplorable state of British agriculture, and possibly to suggest remedies, is to invite criticism, censure and disapproval from all quarters. The land of this country is so varied in character, that conditions are totally different, not only as between county and county, or village and village, but even between neighbouring farms. A tenant farmer, therefore, usually has only intimate knowledge of his own immediate district. Still, there are certain features of the present situation which apply to the whole country, and, in my opinion require the same treatment, so I am rash enough to make an attempt.

English farming has never been solely a business proposition until the present time. In early days the barons held land for power. In later years and right up

271

to the outbreak of the war in 1914, the aristocracy of England held land because of the political and social power which a landlord position gave them, and also because of the pleasures and sports which accompanied its ownership. And that time, in a minor degree, the large tenant farmer rented land for similar reasons, especially the one of sport. Neither the ownership nor the renting of agricultural land was entirely a matter of business. Sport, politics, social considerations, and even religion were much more important features of the rural world than the profit and loss accounts of the actual farming.

The war has changed all that.

To-day the ownership of a rural estate confers few advantages on its possessor. A little social position possibly still lingers, but the old-time benefits are no more. Political power thereby in country districts is a thing of the past. The sporting rights and the stately homes of England are being let to wealthy Americans. The letting of the farms to tenants has become a cold-blooded business proposition only.

And, if that is the position of the rural landlord of a large estate to-day, a similar change has taken place in that of the tenant farmer and owner-occupier. No longer is farming a pleasant, friendly, spacious occupation. No longer can the farmer continue to farm as in the days of his ancestors, in the sure and certain knowledge that with average luck a pleasant jolly life will be his portion, also that, although he will never amass great wealth, he will never know the pinch of poverty and the fear of bankruptcy. The sense of secure well-being is gone. A farm must now pay the interest on the capital involved in it, plus a living to the farmer, in the same way as any other business.

Probably one of the hardest things for farmers to realize to-day that they are considered unimportant people by the majority of the community. When the townsman is hungry the food producer is a very important person, but to-day the consuming public are being fed by foreign countries very cheaply.

As a result of this the farmer has no pride in his occupation. The zest has gone out of farming. For any farmer to go round his fields to-day and view his crops brings him no pleasure. The larger and better the field of wheat, the more useless the whole business appears. He yearns to escape from it, but in most cases he cannot do so: he is caught in a financial cleft stick, and cannot get away. He sends his sons into the police force, into banks, into the Civil Service, anywhere rather than put them into farming. He is hurt and bewildered, and in this frame of mind he snaps at all and sundry, at officials who are sent to worry him by successive governments, at his workmen, and at his family. You may depise him for this, but it is understandable. It is not pleasant for a man to discover that he is engaged in an occupation for which his country has neither use nor interest.

Of course, this business of regarding the other fellow as being of no consequence is not one-sided. Farmers as a class are also rather apt to consider themselves as the sole mainstay of the country, and to regard the town dweller as an unnecessary nuisance.

When travelling by train, on entering one of our large cities, I am always impressed with the endless rows and rows of houses, each with its wireless pole, each with its back yard, each with its brave attempt at some form of garden, and each inhabited by a town household. What does the farmer care about these people's lives and problems? He cares very little, if at all. What do these town workers care about the worries of the British farmer? The answer is precisely the same. They are intensely interested in the price and quality of their food, but hardly at all in the question of from whence it is produced.

And who can blame them? Certainly not the British farmer, for he takes precisely the same view with regard to his purchases, whatever they may be. He cannot supply the consuming public with grain as cheaply as they can buy elsewhere, but he hopes that they will pay him his price, because, forsooth, he is a British farmer! Now what does he do when his own brother, say, offers him feeding stuffs at a higher price than another trader who is no relation? It is conceivable that he may purchase one lot from his brother on the grounds that blood is thicker than water, but does he continue to do so? No, he does exactly what the consuming public do with regard to their food purchases, he buys in the cheapest market.

But in spite of the fact that the farmer to-day is an unwanted individual, he needs to get a living, and that living has, perforce, to come from the buyers of farming products, in other words from the British consumer. Make no mistake about this, it is their money which the farmer wants, and must have, if he is to carry on.

Now in order to obtain anyone's money legally it is necessary to sell them something which they wish to have, either service, knowledge, or goods. The farmer, obviously, must sell goods to the consumer, and the popular idea seems to be that he should sell him wheat.

This idea appears to me to be rather absurd, as wheat is the one commodity produced on the farm which the consumer of food in this country does not want. The unpalatable fact must be admitted that for bread-making, as the public taste goes, we can import a better wheat than we can grow. It is no good saying that the consumer ought to prefer bread made from English wheat. He definitely does not. I know that, as a farmer, I ought to buy English bacon, but I do not do so, because I prefer the flavour of the Danish curing. You may call me a traitor to my calling, if you like, but there it is. In this instance, taste, not price, is the deciding factor, but in almost every case it is on one or both of these two considerations that the purchaser of any article decides. That the buyer's taste may be considered a bad one according to certain standards is immaterial. The fact that it is *his* taste is the important point.

Another fallacy is to talk about the cost of production in relation to the subsequent sale of the article produced. The consumer definitely is not interested in this, neither when he happens to want the article nor when he does not want it. Does a lady worry over the cost of producing a particular hat in a shop window, when she has decided that it is just the one she wants? No, she wants it, and that usually is the reason which decides her purchase.

The best thing is obviously to sell the consumer something which he desires for as high a price as he will pay to satisfy that desire. He does not want the farmer's wheat, but he does want to camp on his land. Charge

275

him for the privilege. He wishes to sit on his own camp stool in a farmer's pasture, and paint a picture. Charge him. Don't worry about the cost of production. He may wish to bathe, to picnic, to play rounders, or to stroll over a field and listen to the song of the lark. Charge him for so doing. He may resent the charge, but if he wants anything badly enough he will pay for it.

Ah, but that is not farming, I hear you say. Granted, but it is something which can be sold to bring in a little profit to the farmer. Wheat-growing may be farming, but does not bring in any profit? The consumer must have milk, but milk on a farm is valueless to him. Put it in a bottle, deliver it to his doorstep, perhaps even open one of his windows which he will leave unlatched for the purpose, and place the bottle of milk between his aspidistra and his geranium, being particularly careful not to wake him up in so doing, and he will pay you. The actual milk represents the smallest part of his purchase; the service, the convenience, and, I think, the not waking him up, count far more.

But many farmers consider that these sort of things are beneath their dignity. They want to farm in the lordly independent style of the days gone by. To-day's prices for grain do not enable them to do so, and they hope that Government aid will be forthcoming to make it possible. What farmers are trying to do, apparently, is to persuade some political party to compel the consuming public either to agree to enhanced bread prices, or to pay the farmer a subsidy or dole from the exchequer. Again it should be noted that one cannot get away from the fact that it is the consumer's money they want.

I am not going into the argument as to whether either

276

or both of these things are just and fair or not. As a farmer I am much more interested in the probability of their happening than in question of fairness. I am convinced that they are not going to happen, and I cannot see that for the farmer to go whimpering to the Government for them is at all dignified.

But nobody seems to be able to get away from wheat as the main theme, when agricultural problems are being discussed. All the various proposals under discussion, no matter to what party programme they belong, are concerned with enabling the British farmer to grow wheat. It seems absurd. It cannot be grown in this country economically, the consumer prefers the imported article, our climate is not suited to its cultivation, and, at the present time, while home-grown grain is only 10·7 per cent of our agricultural produce, wheat represents but a bare 4·3 per cent of our output.

The future prosperity of British farming depends not on government aid in the shape of doles, but chiefly on the ability of the farmer to purchase low-priced imported grain and feeding stuffs for his stock. An honest referendum of farmers to-day would prove this. The dairyman, the stock-raiser, the poultry farmer, the pig-keeper, and the producer of beef or mutton should surely vote for low-priced feeding stuffs. Of course, many of these would not do so if they thought there was a chance of grain-growing becoming profitable once again, because this method of farming is so much more pleasurable to the farmer. I, for one, would much prefer to farm in the style of twenty years ago. But I cannot see that it is likely to happen, neither from an increase in the world's price for grain, nor from any political intervention whereby the community will pay good money in order that I shall have a pleasant occupation.

What, then, should the British farmer do to be saved? The present economic situation is beating him, politicians neither can nor will help him, and the consuming public does not care what happens to him. Some folk talk a lot of the sympathy of the public. If there be such a thing, which I doubt, very certainly *the farmer has not got it*. His only course is to try to force a living out of the public by his own efforts, thereby obtaining in addition their respect, which is of more value than their sympathy. The British consumer is the British farmer's only possible customer, and customers' wishes must be studied.

Now what are the products of British farming, which the consuming public appreciate and prefer to the imported variety? Briefly they are English meat, English dairy produce, and English eggs. These home products are of better quality than their foreign counterpart, and this is the only sound avenue of approach, whether the farmer wishes to get a living unaided or whether he desires to obtain any public interest and assistance in his present plight. The consumer will not pay the farmer more money for home-produced wheat than for imported, but he does and will continue to pay more money for those home products which suit his taste.

Now all these products of British farming, which are superior in quality to the imported variety, are primarily the products of grass farming, and depend largely on cheap imported grain and feeding stuffs. Therefore, generally speaking, the land of England must be put down to grass, if farming is to pay its way. Of course, this idea is anathema to the politicians, to the sportsmen, and also to a large proportion of those immediately connected with agriculture. These folk continue to raise the old cry of wheat-growing, but in their inmost hearts they know that it is economically unsound and unsuited to our weather.

But is the idea of Britain becoming primarily a grass farming country altogether a bad one? Surely it would be using our land in the manner best suited to its climatic conditions, as well as growing those things which the consumer prefers. Generally speaking, for the past hundred years grain-growing has only paid the British farmer when some abnormal condition of things has obtained, such as war or the failure of a large foreign crop. The English climate is far from ideal for grain-growing. It is only on the rare occasions when the price of grain shows a large enough margin over the cost of production to cover the risk of a bad wet harvest, that grain-growing in this country can be considered a justifiable business proposition at all. Many of the old farm agreements have a clause which fines the tenant heavily should he plough any of the existing grass land on his farm, as by so doing he impairs the value of the farm.

But while so much is talked about wheat in these days, the fact that our land and climate are almost ideal for grass farming is rarely mentioned. There is a very small acreage of agricultural land in the whole world so well suited for this purpose as the British Isles. Ireland and the south-west of England have a grazing season of almost the whole year, while the remainder of our country has six or seven months. What other country, save perhaps New Zealand, which is far away from our market, can equal this? Surely the British farmer should exploit the natural advantages of his soil and climate rather than persist in wheat-growing, a branch of farming in which so many other countries beat him, not only in the price at which they can sell wheat but also in the quality of it.

The chief argument against this will doubtless be to point to the over-production of milk in excess of liquid requirements at the present time. The Danish farmer does not worry about over-production in this direction. He produces as much as he is able. The surplus to his country's liquid requirements is made into butter and cheese, and the skim milk, whey, and other by-products of their manufacture are returned to the farmer and fed to pigs. The dairy products thus made and also the bacon thus produced, he sells chiefly to us. In this respect cheap imported grain is one of his chief assets.

The various advocates of wheat-growing in this country try to evade the real issue of its economic unsoundness by saying that it is a bulwark of national defence. Well, even if one agrees with this statement, the unpleasant fact remains that the nation seems very reluctant to pay for this national bulwark, and very certainly farmers cannot pay for it any longer.

But there is another view. Wheat-growing to-day is the reverse of a bulwark. The present price for wheat does not provide a sufficient return to the farmer, to enable him to pay his way or to keep his land in good heart. In order to meet the deficit in farming accounts, many farmers increase the annual acreage of their corn crops, thereby impoverishing their land. Apart from the low prices, they find that they are getting diminishing yields year after year, and the national bulwark (?) is therefore getting weaker and weaker. Under intensive stock farming the fertility of England's soil would be enhanced greatly, and should the time ever arise when grain-growing would be a national bulwark, we should have a strong one on which to depend. The danger that as this fertile grassland could not produce grain in less than ten months, the nation might be short of food in war time if imported supplies were stopped, could be obviated by a national storage of a six months' supply, since wheat will keep.

Doubtless many people will disagree with the idea that Britain should abandon wheat-growing and become primarily a grass farming country. But that his change is taking place all over our land cannot be denied, neither can it be stopped. The politicians, the press, and and the farmers, who are endeavouring to stop it, will soon discover that they are but modern Canutes. The tide of economic forces cannot be stayed. Even if grain-growing be bolstered up by political action it can only last a very short time, because it is economically un-sound, unsuited to our climatic conditions, and also to the consumer's taste as compared with meat and milk farming.

The foregoing has been written in the assumption that any proposed Government assistance to agriculture is almost certain to be in connection with wheat-growing and as such, bound to fail. Also, as most townsfolk deprecate the idea of any assistance from the National Exchequer being given to farmers, they are almost certain to vote against it. But if there is any national desire that our farming industry be put upon a prosperous basis, surely some political party should, at least, investigate the problem with the idea of helping the farmers to make this drastic change in their methods, which is happening all over the country.

That it is happening there can be no doubt. More and more land is being sown down to pasture each succeeding year. Wheat-growing has always been considered as the sheet anchor of British farming, and most of our farms, farming agreements, and customs have been planned for that purpose. The change-over to intensive stock farming as the principal feature, while it is taking place, is doing so in a very painful manner to those most concerned.

Many an arable farmer, having arrived at the point in his farming career where he must change over to grass or quit, is unable to do so for financial reasons. A few hundred pounds worth of arable implements enable him to plant and reap his corn, thus getting deeper and deeper into the mire each year. To grass, fence, and water his arable land will take all his money. What is the good of doing this if he is unable to find the capital to stock his grassed farm? Those who have a sufficiency of capital are making, or have already made, this change in their farming methods. The others, who have not the wherewithal, continue to grow grain in the vain hope that something miraculous may happen.

Is it possible for government aid to make this change-over in farming methods less painful, or must many good arable farmers go down and out, leaving their farms to newcomers who have the capital to accomplish this alteration?

EPILOGUE

One of the chief arguments in favour of a guaranteed price for wheat is that it would cause a lot of land which has been grassed down in recent years to be brought into cultivation once more. But would it? The usual figure quoted is a minimum of about fifty shillings per quarter for wheat. That price might prevent any more arable land being sown down, but very little grassland would be broken up in consequence. It costs a lot of money to grass, fence, and water arable land. Wheat at fifty shillings would not tempt many farmers who have sunk a considerable amount of capital in changing over their farms from arable to grass, to scrap this expensive improvement. I have grassed the whole of my farm, and I might be tempted to change my farming system for wheat-growing at seventy-five shillings per quarter. That is the minimum at which I should consider even the purchase of arable implements, and the consequent scrapping of my new pastures and fences. And that, as Euclid says, is absurd, while to make this change, relying on the stability of any political action, would be ridiculous.

Ours is a thickly populated country. The motor car makes it possible for the land around each town to be farmed by producer retailers of those farming products which the town's population desire. Possibly there are certain tracts of flat good arable land in our country where grain might still be grown profitably by mechanical cultivation, being alternated with sugar beet in districts suitable for the growing of this crop, and in which sugar factories are situated. Farther away from towns, stock-raising, milk-production for the London market, and the production of grass-fed beef and

283

mutton could be carried on under almost colonial conditions. Any development of the Empire Free Trade idea would doubtless give us a sufficiency of grain for our needs, but in a great measure we are to-day importing a considerable quantity of meat and dairy produce from countries outside the Empire.

Besides, this suggested meat and milk farming, if handled nationally, need not mean a reduction of labour to any great extent. Each farm, though chiefly grass, would grow a certain amount of grain for the production of feed litter, and thatch for the farm needs. This grain would not be grown for sale, but only when the economic needs of the particular farm made it a paying proposition. Generally speaking, this type of farming would require a lot of intelligent and highly paid labour.

It is difficult for farmers to view this question dispassionately—I find it so myself. In spite of all my efforts and intentions to the contrary the idea that the farmer personally should receive some form of government aid, has crept into this last chapter. I consider that the farmers of our country were treated unfairly some years ago by the repeal of that portion of the Corn Production Act which gave them some stability of prices, while the other part relating to their production costs remained unchanged, and farmers generally hold this opinion.

But can anything in this life be termed fair? Nature, health, sickness, are these fair? While injustice can be, and often is, created by Act of Parliament, I do not think it either possible or practicable to make things fair by the same means. Any attempt to do so usually creates another and often worse injustice. Most of the difficulties and handicaps under which farming has to be carried on to-day, are the direct result of attempts to make things fair for the consumer or labourer. In consequence, the farmer considers that he should be so treated to redress the balance, and pleads for some political action in his favour.

In those last three words lie his chief mistake. While the nation does consider the interests of the labourer and consumer, and may possibly consider that the farming industry is worthy of concern in its present condition, the personal plight of the individual farmer does not worry it at all. That many farmers will fail and go under is to the nation of no consequence. That a large national industry is going downhill is another matter. The safeguarding duties in other trades were not imposed to help the employers but to aid the particular industry, solely on the grounds that the nation could not afford to see it fail. It may be said that this attitude is unfair, and possibly unwise, but it cannot be said that it is untrue.

That is the position which farmers must face. They will get no financial aid from any government for themselves, neither will they obtain any for the farming industry unless that industry produces those things which the population of our country desires.

In any case, if any lasting good to our country is to come from any political intervention in the present agricultural situation, it must be on these lines, and what is perhaps much more important, there must be considerable more honesty on both sides when these problems are discussed. The farmer must realize that the town consumer, in normal times, is never again going to pay a sufficiently high price for farming products to enable the producer to live in the style of 1906–1921. The politician must admit that still higher wages and still shorter hours for the agricultural worker can never go hand in hand for long with a cheap breakfast table for the town dweller. There can be no brief for political action to establish wheat-growing, but the city voter may consider that home-produced meat and dairy produce are worth having on the grounds that they are produced in conformity with known standards of cleanliness and quality.

Given such honesty on both sides there is surely the possibility of a prosperous British agriculture such as I have outlined.

That there is one man in the farmer's councils who possesses such honesty, I know. He is blessed with sufficient integrity, breadth of vision, and capacity, and, above all, with backbone enough to pursue his way in spite of any taunts and insults from those whom he represents. Whether there is a politician in our land to-day who is so equipped, I can only wonder.

I do not expect many people to agree with the arguments I have advanced in this concluding chapter, but I would suggest that those who study it fairly will find a grain or two of wheat amongst the chaff. Should the perusal of it cause anyone to investigate the farming problem from this angle, the writing of this last chapter will be amply justified.

July 1931.

Since the writing of the preceding chapter, a great

EPILOGUE

change has come over the political situation, and it looks as if some genuine attempt by parliament to help our agricultural industry is probable in the near future.

Must this be solely on a wheat basis as before? Will they make another attempt to put the clock back?

English farming is definitely turning away from grain-growing. Why should not this parliament give it a helping hand in that direction, which has been proved by the last seven years of depression to be the right one?

November 1931

Printed in Great Britain
by Amazon.co.uk, Ltd.,
Marston Gate.